Boiler Plant
and
Distribution System Optimization
Manual

by:
Harry Taplin, P.E.

Published by
THE FAIRMONT PRESS, INC.
700 Indian Trail
Lilburn, GA 30247

Library of Congress Cataloging-in-Publication Data

Taplin, Harry, 1935-
 Boiler plant and distribution system optimization manual /
by Harry Taplin.
  p. cm.
  Includes index.
  ISBN 0-88173-142-0
  1. Steam-boilers--Efficiency. 2. Steam power-plants--Design and
construction. I. Title.
TJ288.T36   1991   621.1'83-dc20   91-3585
                                                          CIP

Published by The Fairmont Press, Inc.
700 Indian Trail
Lilburn, GA 30247

Printed in the United States of America

10 9 8 7 6 5 4 3 2 1

ISBN   0-88173-142-0   FP

ISBN   0-13-108110-5   PH

The statements, conclusions and suggestions in this manual are intended to inform. They are meant in no way to be applied without the judgment of responsible and experienced engineers or qualified energy managers. Procedures and techniques in this manual are intended as guidelines and must be judged for applicability by consulting actual conditions, safety, cross influences with other systems and components and applicable laws and engineering standards. While every effort is made to provide dependable information, the publisher, authors, and editors cannot be held responsible for any errors or omissions.

Distributed by Prentice-Hall, Inc.
A Simon & Schuster Company
Englewood Cliffs, NJ 07632

Prentice-Hall International (UK) Limited, London
Prentice-Hall of Australia Pty. Limited, Sydney
Prentice-Hall Canada Inc., Toronto
Prentice-Hall Hispanoamericana, S.A., Mexico
Prentice-Hall of India Private Limited, New Delhi
Prentice-Hall of Japan, Inc., Tokyo
Simon & Schuster Asia Pte. Ltd., Singapore
Editora Prentice-Hall do Brasil, Ltda., Rio de Janeiro

Table of Contents

Table of Contents (Continued)

Table of Contents (Continued)

Table of Contents (Continued)

Table of Contents (Continued)

Table of Contents (Continued)

Table of Contents (Continued)

Table of Contents (Continued)

Table of Contents (Continued)

Acknowledgements

This book evolved from the Association of Energy Engineers Boiler Optimization Course and other similar courses which have been ongoing for more than ten years. It was written in response to the need to manage energy as efficiently as possible, to reduce the growing smoke stack pollution problem and to cut back on the release of carbon dioxide to the atmosphere which is reported to be the cause of the greenhouse effect and global climatic changes. Many boilers, furnaces and other combustion processes can be made more efficient. This in turn can reduce pollution and save fuel and money.

The sources of information for this book were many and varied: The Association of Energy Engineers, the American Society of Mechanical Engineers, The American Boiler Manufacturers Association, the U.S. Department of Energy and Research Institutions such as the Navy Civil Engineering Laboratory and Brookhaven National Laboratory all contributed the material in this book. Many boiler, instrument, control system and burner manufacturers also contributed in one way or another.

I would like to thank Mr. L. Ron Hubbard for providing valuable information on communication and education, which helped me to organize, understand and communicate this complex subject.

There are many talented, resourceful and conscious people involved in the boiler and related industries who have done a great deal to advance our civilization to its modern level. Many of them have attended my courses and shared their insights and problems with me. I would like to thank them all for their support and for sharing their concerns and discoveries, which have contributed in many ways to this book.

Harry R. Taplin, Jr., P.E., CEM

Introduction

This book has been designed to make the job of boiler plant optimization at your facility simple and understandable. This book was written for the plant manager, boiler design engineer, energy engineer, plant operators, troubleshooters and anyone else interested in improving combustion efficiency.

One of the most productive ways improve efficiency and profits is to expertly manage the energy costs of utilities, especially in the cost of boiler fuel. Managing an efficient plant involves a different viewpoint from normal day to day plant operations. Unfortunately, operational challenges seem more urgent than managing energy efficiently. One old timer put it well when he said, "improving plant efficiency is not a necessity, but keeping it running is a necessity."

Most plant personnel attend to the important job of running a dependable plant. They might need help seeing the challenge from the viewpoint of saving energy and reducing fuel expenses. That's what this book is about.

Why is this so important? The answer lies in the fact that your company has to earn the money it pays for fuel from the sale of some product or service. There are large expenses involved in production of these products or services and it makes no sense to waste these efforts in the boiler plant with huge hidden costs from wasted energy.

Lets look at the situation from another viewpoint. Suppose the net profit of your operation is 10% and that when you have followed some of the suggestions in this book, you discover you can save your company more than $250,000 a year in reduced fuel costs. How much is this really worth? Well, it's about 2.5 million dollars. There are a lot of people who would be handsomely rewarded if they could boost annual sales by 2.5 million dollars. When you produce "cost avoidance" by increasing the productivity and efficiency of your plant you are doing the same thing when the bottom line is examined. These profits from avoided costs keep your company in business and assures that you will have your job into the future. You may have more to do with the success of your company than you have been lead to believe.

How do you get this job done? It seems like it might be a highly technical matter which demands a tremendous amount of time, knowledge and experience with boiler plants and distribution systems. Actually, if this challenge is approached properly, it is not complex or difficult.

Harry R. Taplin, Jr., P.E., CEM

Chapter 1

Optimizing Boiler Plants
Establishing the Ideal Scene

Optimizing the performance of boiler plants and distribution systems is a broad and rewarding subject. On the surface you might think it requires a great deal of knowledge and insight. It covers many engineering subjects and involves an understanding of the dynamics of operating systems composed of many subsystems and components. If the job is approached on a systematic basis, it is much simpler and you will be more successful.

As-found efficiency is the efficiency for a boiler in its existing state of repair and maintenance. This efficiency will be used as the baseline for any later efficiency improvements. It is very important that the as-found efficiency be recorded because it will serve as a benchmark to estimate the value of a Boiler Optimization Program.

The beginning point for Boiler Plant Optimization Program is to establish the *As-found efficiency* for each boiler and system. This information will serve as a datum to show the expected benefits of the optimizing program.

It may be that millions of dollars can be saved by improving plant and distribution system efficiency. The *as-found* conditions will serve as a datum for establishing this fact and also show the folly of neglecting boilers and their energy distribution systems.

Also, if the plant is found to be efficient, it will bring credit to those responsible for the plant. In any case the *as-found efficiency* is important for economic evaluations and justifying additional personnel and modifications to existing systems.

The next step will be to tune-up the boiler and accomplish any maintenance and repair identified during the initial testing.

Tuned-up efficiency is the efficiency after operating adjustments, lowering excess air, and minor repairs have been completed. This will be the baseline efficiency for estimating all future savings.

There are many minor problems which can develop during the life of a boiler and distribution systems that cause non optimum performance. Over time, they can waste a great deal of energy.

To make an honest and accurate evaluation of the potential savings available, all equipment must be in a normal state of repair with the boiler air/fuel ratios at designed levels. If this is not the case, estimates of savings and justifications for new equipment and modifications will contain false information leading to poor decisions.

It is very important that the tuned-up efficiency be accurate, because the economic

1

benefits of improvement options will be estimated from this efficiency.

Next, there should be an estimate of the savings if the efficiency were to be improved to a reasonable level. This information is used to determine if further work on a boiler is justified and to set priorities for competing projects.

**Maximum economically achievable efficiency** _is the efficiency that can be achieved, with efficiency improvement equipment added_ only _if it is economically justifiable._ **Table 1.1** _lists this efficiency level for a wide range of boilers and fuels._

It might be necessary to know what the maximum efficiency of a boiler could be if the cost of modifications were not considered. For example, knowing maximum efficiency would be helpful if fuel were very scarce.

**Maximum attainable efficiency** _is the result of adding the best available efficiency improvement equipment, regardless of cost considerations._

Also, it might be cheaper to raise efficiency (productivity) of a boiler rather than install another boiler, at high cost, to keep up with growing plant steam demands.

Table 1.2 shows _maximum attainable efficiency_ for various size boilers and fuels.

The First Step.

The first step is to find the _as-found efficiency_ of each boiler.

Efficiency improvement potential is based

on the _as-found efficiency_. For example, it can be used to show the value of testing the performance of boilers on a regular basis by identifying losses caused by the drop in efficiency from the well-maintained or tuned-up condition.

It can also indicate if your maintenance program needs improvement and will show the dollar losses if efficiencies are allowed to drop.

An efficiency monitoring system is also a good way to check the work of contractors and engineering consultants to insure they actually improved efficiency or to find out if they created new problems with your equipment.

When one speaks of boiler efficiency, a degree of generality is present unless the term efficiency is further defined. Boilers normally operate over a range of efficiency **(Fig 1.1 through 1.3)**.

The efficiency of a boiler usually falls off as the bottom end of the of the turn-down ratio is approached. This is because the volume of air through the burner is greatly reduced affecting the performance of air-fuel mixing. The excess air must be increased to compensate for this problem to prevent smoking and incomplete combustion which could lead lower efficiencies.

As the boiler approaches its maximum firing rate, its capability to recover all of the heat from the combustion process diminishes and stack temperature rises. This is the primary reason for the fall off of efficiency as the maximum firing rate is approached.

Optimizing Boiler Plants
Establishing the Ideal Scene

Maximum Economically Achievable Efficiency Levels

Fuel	Rated Capacity Million BTU's/HR		
	10 -16	16 - 100	100 - 250
Gas	80.1%	81.7%	84.0%
Oil	84.1%	86.7%	88.3%
Coal Stoker	81.6%	83.9%	85.5%
Pulverized	83.3%	86.8%	88.8%

Table 1.1

Maximum Attainable Efficiency Levels

Fuel	Rated Capacity Million BTU's/HR		
	10 -16	16 - 100	100 - 250
Gas	85.6%	86.2%	86.5%
Oil	88.8%	89.4%	89.7%
Coal Stoker	86.4%	87.0%	87.3%
Pulverized	89.5%	90.1%	90.4%

Table 1.2

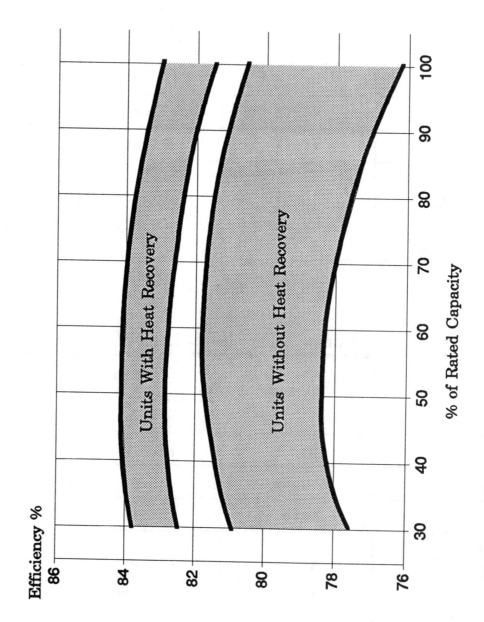

Figure 1.1 Typical performance of Gas Fired Water Tube Boilers

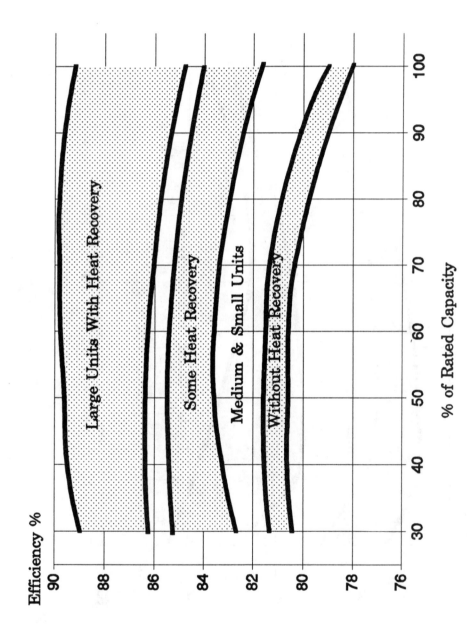

Figure 1.2 Typical performance of oil fired water tube boilers.

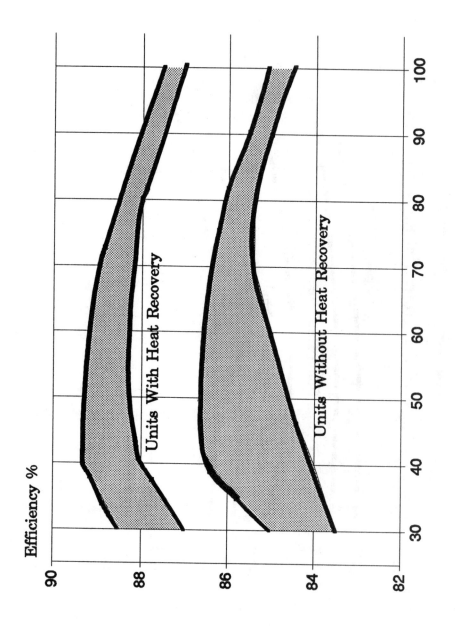

Figure 1.3 Typical performance of Pulverized coal fired boilers.

Due to the highly diverse range of boiler designs over many years, any curve of performance which might be presented can only be considered for general illustration. **Figures 1.1 through 1.3** are the result of an extensive survey sponsored by the Department of Energy and are to a large degree typical for the aggregate boiler population in the U.S.

A curve of efficiency over the entire load range should be developed for each specific boiler. This curve can serve as an accurate statement of a boilers efficiency, but further examination will be necessary to establish its real operating efficiency. **Figures 1.4 through 1.6** show typical characteristics that may be found when plotting these curves.

Where does the boiler operate typically? Does it stay at full load or some other fixed load most of the time? Does the load vary quite a bit, just what is the average load and average efficiency for a particular boiler? Does it cycle on and off a lot? The more accurate the data, the closer the estimation of efficiency to actual conditions.

These factors argue against arriving at boiler efficiency easily or quickly using the heat loss method. This being the case, judgments and estimates are usually applied unless comprehensive metering exists for accurate input-output efficiency monitoring.

Loss prevention

A further refinement of boiler optimization comes under the heading of **"Loss-Prevention"** which is a system of identifying key performance parameters and identifying them with annual fuel or dollar losses when they vary from the ideal.

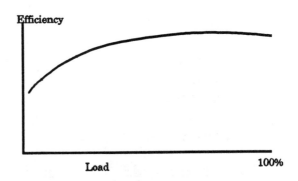

Figure 1.4, efficiency curve showing high excess air losses below half load.

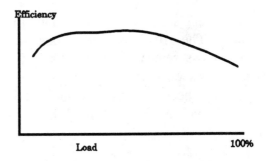

Figure 1.5, efficiency curve showing high stack temperature losses above half load.

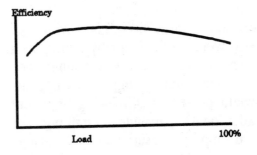

Figure 1.6, a typical efficiency curve showing moderate excess air losses at low firing rates and heat exchange losses at high firing rates.

Optimizing Boiler Plants
Establishing the Ideal Scene

Once the ideal scene has been defined, loss prevention can be indexed to these and other key indicators:

a. Stack temperature
b. Excess air
c. Percent make up water
d. Blowdown losses
e. Condensate return temperature

A loss prevention program is an essential part of boiler plant optimization as it deals with the control of specific elements of boiler and system operation and maintenance.

Thus, two ideal scenes are appropriate to any plant. The first is ideal design. Efficiency measurements can be used to justify projects such as economizers, oxygen trim systems, heat recovery options and such to achieve the ideal design for the plant.

The second ideal scene deals with how the plant is operated and maintained relative to unnecessary energy losses within the control of plant personnel.

Optimizing boiler management

If tests show an annual fuel savings of $100,000 were possible just by tuning up a boiler and correcting simple maintenance problems uncovered during a tune up, this fact would present a convincing argument for additional personnel, training or technical support to insure that these savings are maintained. Hidden costs from efficiency problems can be a very expensive oversight.

The term management is not often discussed in connection with boilers and distribution systems, probably because they are operated, but not managed in the sense of maintaining optimum performance.

A good many managers feel that the efficiency of a plant should never change. What is missed is that they are a very dynamic energy consuming monster that must be managed. If not prevented from wasting fuel, they will surely be feasting on your utility dollars.

Plant engineering and maintenance staff are often too busy or undermanned or don't understand the magnitude of losses involved in boiler plants and distribution systems.

Too often plants don't have instrumentation to measure what is happening with their plant and distribution systems.

Accountability

If you want to be successful in an energy conservation program, accountability must play a key role, there must be someone who is accountable for energy. This is just not someone who checks the utility bills. It may take a professional who can spot problems and knows how to correct them.

Surveys of many plants have shown very conclusively that there is usually a lot more attention on "operating" than on "managing" in most plants.

This is clearly evident from examining plant instrumentation, most of it is for plant operations. There usually aren't any instruments or means to measure stack losses, blowdown losses, condensate system losses or to measure over-all system efficiency. Existing instrumentation is used for estimating these values, however typical instrumentation errors can produce large errors.

Measure to Manage

The **2M** system **"Measure to Manage"** has been a very successful approach to reducing plant energy costs. It is too easy to waste energy without it.

There is usually no accountability system except in the some office where the utility bills get checked and paid. Here, the math of the bills is reviewed and perhaps comparisons of last year's bills to this year's are made. Plant personnel are often left out of the loop and this doesn't create any incentive to manage efficiently or to save energy.

Savings of more than 20% have been achieved just by installing a system to measure and manage energy in a plant to find out what was happening.

Once you have complete and accurate data, it's easy to see what is happening in any plant or distribution system. Establishing a system to get data, interpret it and then controlling unnecessary losses is the real challenge.

An effective energy conservation program

The essential elements of an effective energy conservation program are:

1. Management commitment
2. Boiler and system testing
3. Economic evaluation of energy conservation projects
4. Assignment of project priorities
5. Final plan
6. Plan implementation
7. Loss control management program
8. Monitoring results

1. Management Commitment

One of the most important aspects of an energy conservation program, often missing, is management support. If nobody seems to be interested in the efficiency or productivity, things can get very bad.

It was just a few years ago when a vice president of a major U.S. corporation addressed a group of energy engineers at a national conference after receiving an award for being "The Energy Manager of the Year." He stated that the most important contributing fact to his very successful energy conservation program was the full backing of the president of his company. He said, "If the boss shows an interest, something is going to get done."

2. Boiler and System Testing

In the first place, the plant or system may not have been designed for optimum efficiency. Design omissions, deviations from plans by contractors, modifications over the years, changes in operational requirements and maintenance modifications may have all influenced efficiency. It is wise not to make any assumptions about performance. Testing and actual measurements will tell the story about how the system is performing now.

3. Economic Evaluation of Energy Conservation Projects

Testing and evaluation will form a good basis for judging the opportunities for saving energy with the boiler or system as it operates in "todays" environment. Payback period, return-on-investment, life cycle costing and other economic evaluation methods are important measurements of the real worth of a project to modify a boiler or improve system performance. When faced with many competing projects, economic evaluation will show which is best and which projects won't pay.

4. Assignment of Project Priorities

There is usually competition for limited funding to improve efficiency and save energy. Management must have data on which choices will be most beneficial. The economic evaluation process makes it possible to rank various projects on the strength of their merits.

5. Final Plan

A wise man once said "If you don't have a plan, then you plan to fail." A final plan infers that all facts have been considered and a smart and orderly course will be followed.

6. Plan Implementation

The plan has now become a working tool. Like many other tools, plan implementation will get the job done.

7. Loss Control Management

Assign an annual loss value in dollars to indicators of system performance. Each operating parameter that will have a significant impact on operating cost and energy losses should be monitored.

Possible items for loss control management are: (a) for each 10°F rise in stack temperature there will be a $10,000 annual loss if not corrected, (b) for each 10% change in excess air fuel costs will go up $15,000 annually (c) For each 10°F the condensate returns below the ideal temperature, $40,000 will be lost in a year and so on. Each plant will have its own values.

8. Monitoring Results

The use of computers and the ease of acquiring data from plant operations provides new opportunities to actually monitor performance on a full time basis rather than guessing or making assumptions. It's easy to claim that equipment will improve performance or repairs were made properly, a monitoring system will tell the story and can be a valuable tool in managing an efficient plant.

In the following chapters we will go through a step by step program for a typical facility where you will be shown how to identify and deal with the many problem areas and establish the "ideal scene" for your plant.

Chapter 2

Basic Boiler Plant Efficiency

This chapter is designed to show the logical steps that should be followed to find basic boiler plant efficiency. It covers the basic steps for plant survey and testing to identify sources of losses in the plant and distribution system.

Fuel Consumption

The first requirement is to gather accurate fuel use information for each boiler. **Figure 2.1** shows the basic information required.

Each boiler should have a separate fuel meter so fuel use can be accurately tracked for each boiler. The 178 gal/hr and 25 million Btu/hr represent the maximum input for this boiler.

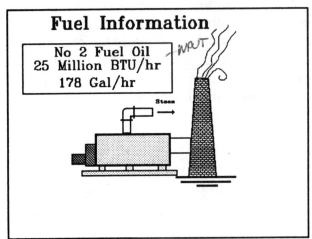

Figure 2.1 Fuel use information.

Fuel Cost

Figure 2.2 shows the basic data on fuel costs necessary for financial evaluations. This information is important for the analysis of dollar losses and for evaluating the dollar value of fuel saving opportunities.

The 150 dollars per hour represents the maximum firing rate.

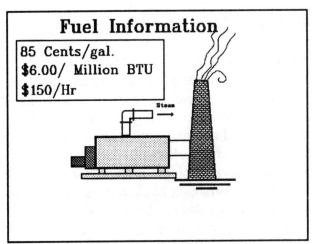

Figure 2.2 Fuel cost information.

Annual Fuel Costs

Figure 2.3 gives annual fuel use information for our typical boiler, this data is essential for evaluating potential savings and ranking energy conservation options based on annual fuel consumption. It will also provide data for evaluating the effectiveness of energy conservation options after they have been installed.

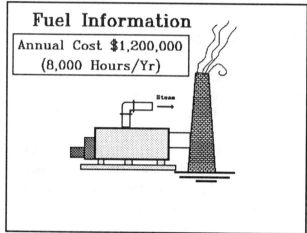

Figure 2.3 Annual fuel costs.

Analyze Performance

Analysis of plant performance requires identification of the principal losses. Stack Losses are usually the largest single loss in boiler operations. By testing the exit gas temperature and excess air levels, the sensible and latent heat being lost up the stack can be established. A rough estimate of boiler efficiency is often obtained by subtracting stack losses from 100 percent. It must be noted that many other losses are involved in operating boilers and this method of using just one loss for the efficiency determination has limited value. There are a number of very good portable analyzers on the market that give stack loss efficiency directly, they also show other important information like excess air, stack temperature, carbon monoxide, combustibles and oxygen readings that can be very useful in identifying the sources of boiler problems. **Chapters 5 and 6** cover how to measure these losses in more detail.

Stack Losses

As shown in **Figure 2.4**, stack losses are 22.8% which provides a rough efficiency of 77.2%.

Stack loss is not the only operational boiler loss. Because it is the largest, using this value provides rough information on efficiency. It involves two very important parameters, excess air and stack temperature which are items that should be given first priority in any boiler optimization program.

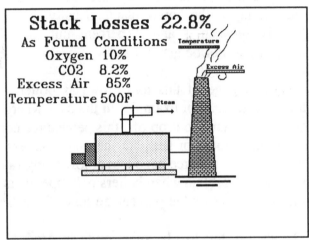

Figure 2.4 Stack losses.

Blowdown Losses

Blowdown losses are often overlooked because they are hard to measure and water chemistry work is not always fully understood by plant personnel. As shown in **Figure 2.5**. this boiler has a blowdown rate of 11.4% of its total steam output. Hot water is being dumped into a drain or sewer system and excess heat is being vented to the atmosphere as flash steam. This amounts to 2,280 LB/Hr of <u>very hot</u> water which is responsible for an energy loss of 4.6%

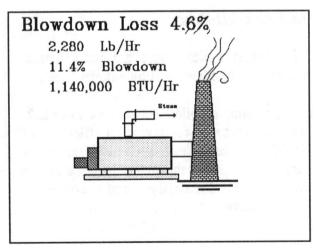

Figure 2.5 Blowdown losses.

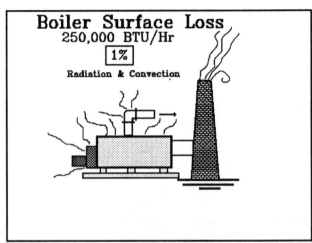

Figure 2.6 Boiler surface losses.

Surface Losses

Another significant loss is the boiler surface loss (**Fig 2.6**) caused by radiation and convection from the boiler surface and valves and piping in the boiler plant. Usually this loss is between 1 and 3 percent but this number can not be generalized.

The percentage of this loss grows as boiler load decreases. For example, in the case where this loss is 2% of the full firing rate. When the boiler load starts to drop off, this percentage continues to go up because the boiler is still losing the same amount of heat but the output is less so the loss is a bigger percentage of the firing rate. The actual percentage loss at low firing rates could be 8 to 10 percent. This becomes a real significant problem when boilers only operate at a small fraction of their design output capacity. **Figure 2.6** shows the percentage loss of this boiler at 100% of its full rated output.

The level of this loss becomes quite apparent when one enters a boiler room, some are very hot and uncomfortable. Good insulation, design and maintenance can reduce this loss significantly.

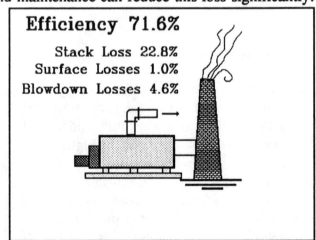

Figure 2.7 As-found efficiency.

As-Found Efficiency

The largest boiler losses are stack losses, boiler surface losses and blowdown losses.

In our example boiler, shown in **Figure 2.7**, they add up to 28.4% giving an efficiency of 71.6%. These losses are not all of the losses but they represent the major losses that can be measured easily and corrected economically.

Tune-Up

Figure 2.8 shows the first step in dealing with boiler losses. The Tune-Up is almost always found to be necessary because is common to hide problems and defects in the burner system, control system and other components by increasing excess air.

When a boiler is having problems it usually begins to smoke. Problems have often been "fixed" by increasing excess air. If no one is monitoring and correcting the excess air level, this situation can cost a lot of wasted fuel. Be advised that problems that have been "fixed" by increasing excess air will reappear when a tune-up is attempted.

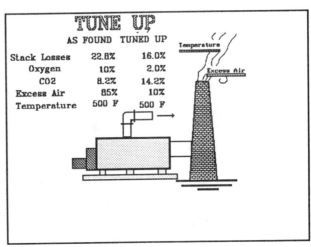

Figure 2.8 Boiler tune-up.

By reducing excess air from 85% to 10% in the example boiler, fuel losses have been reduced by 8.7%. This is a typical fuel savings for many boilers.

Reducing Stack Temperature

Figure 2.9 shows that a fuel savings of 2.73% can be produced by reducing stack temperature from 500 F to 390 F. This high as-found temperature may be caused by soot or scale deposits in the boiler or by hot gas short circuits through defective baffles.

Economizers and air preheaters are also used to reduce the net stack temperature.

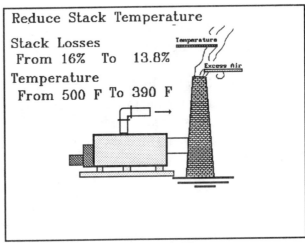

Figure 2.9 Reducing Stack Temperature.

Blowdown Heat Recovery

Figure 2.10 shows that by recovering energy from the blowdown system and heating the incoming makeup feed water with this recovered energy, losses can be cut from 4.6% to 0.6%. Heat recovery equipment in this case can reduce blowdown losses from 500 BTU/lb to 66 BTU/lb.

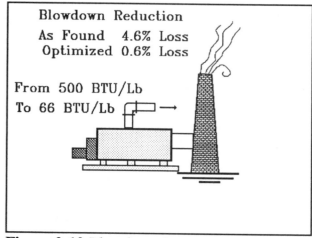

Figure 2.10 Blowdown heat recovery.

Optimized Efficiency

Figure 2.11 shows that by taking the standard actions of tuning up the boiler, recovering blowdown losses and reducing stack temperature, efficiency has been increased from 71.6% to 84.6%

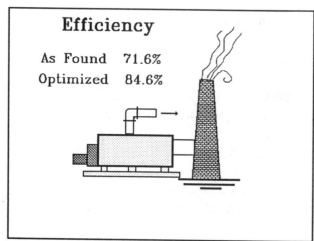

Figure 2.11 Optimized efficiency.

Cost of Steam

Figure 2.12 shows one of the benefits of improving efficiency. The cost of steam has been reduced from $8.38 to $7.09 per pound.

Figure 2.12 Change in steam cost resulting from boiler plant optimization.

Boiler Productivity

Figure 2.13 shows how the productivity of the boiler has been increased from the "as found" output of 17.9 million BTU/hr to an optimized 21.4 million BTU/hr, an increase of 20%. This is also an increase from 17.9 to 21.4 thousands of pounds of steam per hour.

Increased boiler productivity can save money in two important ways; first if additional boilers are being kept on the line at reduced firing rates to pick up demand surges, this additional capacity may provide a safety

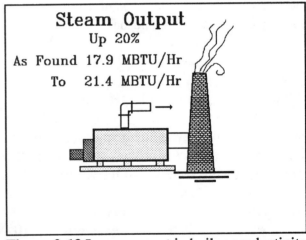

Figure 2.13 Improvement in boiler productivity from boiler plant optimization.

margin which will allow the extra boilers to be shut down. If there is a need for more capacity because of plant expansion or production requirements, increasing the efficiency and productivity of existing boilers may eliminate the need for high capital investment and pollution compliance costs for new boilers.

Dollar Savings

Figure 2.14 shows the dollar savings for this boiler are quite impressive.

Total savings are **$193,574** a year. The annual fuel bill will shrink from $1,200,000 to slightly over a million dollars.

The tune up will save **$104,082**, the lowered stack temperature **$32,754** and the reduced blowdown **$56,738**. Each plant is different but this example is not unusual.

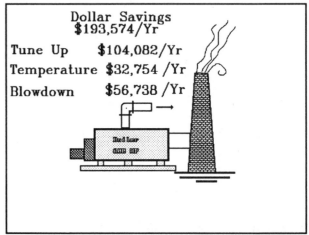

Figure 2.14 Dollar Savings from boiler plant optimization.

What is the real value of fuel savings?

Every fuel dollar saved can be put in the profit column for your company. This means that fuel dollars saved are actually worth a lot more than the plant savings listed above because of the costs involved in earning the money to purchase fuel in the first place are avoided.

For example if a company is earning a profit of 10% on its product, each $1,000 saved in boiler fuel actually avoids business expenses of $10,000 to earn the money for the fuel in the first place.

In our example we have an actual savings of slightly over $100,000 and the company did not have to do one million dollars in business to earn the money to pay for the fuel. This savings amounts to <u>profit</u>.

Chapter 3

Distribution System Efficiency

Steam System Losses

The steam system is ideal for moving energy through piping. No pumps are needed because when the latent energy in the steam is expended it condenses to approximately 1/700 of its original volume causing a significant pressure drop. Replacement steam flows to fill the volume of the steam which has condensed, often reaching velocities of 8,000 to 12,000 feet per minute on the way.

The "pure" water from the condensed steam still contains valuable energy and for efficient operation it should be recovered and transported back the boiler plant where it is reheated into turned back into steam and sent out into the system once again.

Boilers and their distribution systems have a closely interrelated efficiency. The losses involved in distribution systems have a significant impact on boiler operations and efficiency. For example, if the hot condensate from the recovery system does not return to the boiler plant it must be replaced with cold makeup water which must be heated and also receive chemical treatment. Every Btu wasted in the distribution system must be replaced by the boiler which wastes additional energy. This cycle feeds on itself.

When energy losses in the steam distribution system are corrected, boiler plant losses are reduced too.

The steam distribution system moves the steam from the boiler to perform work at some distant location. They do not usually exist separately and they work together to establish a combined efficiency.

The distribution system for our example will be a typical 8,000 foot system with 2,000 feet of inactive piping. The system pressure is 250 PSI, the same as the boiler example covered earlier in Chapter 2. **Figure 3.1** shows a typical steam distribution and condensate recovery system which will serve for our example.

The losses to be investigated are:

1. Insulation losses

2. Condensate system losses

3. Trap flash losses

4. Steam trap leakage

5. External steam leaks

6. Internal steam leaks

7. Inactive steam piping

The steam load on the system in this case is 17,900 pounds of steam per hour, the same as the as-found conditions in the boiler example in Chapter 2. Total energy being delivered to the distribution system is 17.9

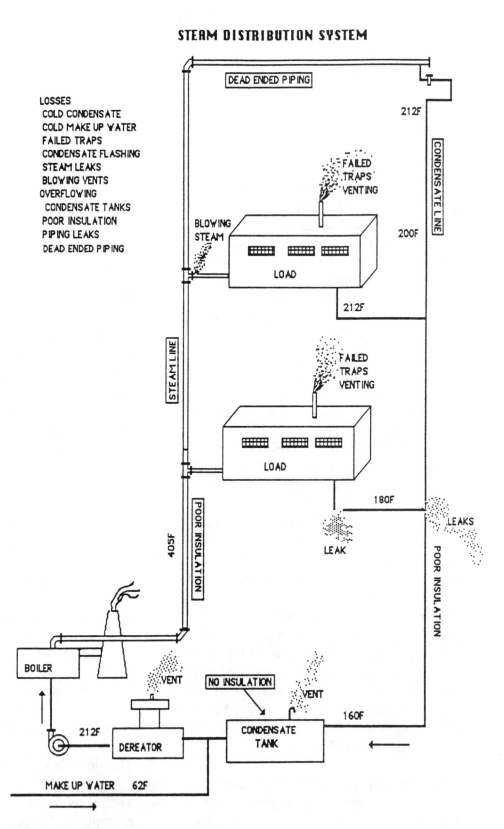

Figure 3.1 A typical distribution system showing how energy is lost.

million Btus per hour. The cost of steam is $8.38 per million Btus.

Each pound of steam lost from the distribution system has an enthalpy of 1140 Btu/lb. The make up water for this system is 42 F, containing 10 Btus per pound. The energy value for the steam is actually 1150 Btu/lb. When we subtract the heat value of our "heat-sink", the 42 F make up water, the value is 1140 Btu/lb.

Insulation Losses

Any section of hot pipe, valve bodies, unions, flanges, and mechanical steam traps such as float and thermostatic, inverted bucket and the bodies of disc traps should be insulated. Insulation will pay for itself quickly as well as prevent the overheating of working spaces and the elimination of hot spots which can cause burns.

Steam users are fully aware of the need to insulate hot surfaces but it takes constant vigilance to keep steam systems in good condition.

An Easy Way to Measure Insulation Losses

An alternate way to measure insulation losses is to weigh the condensate draining from the section of piping under study. If this piping is typical of the entire system, then an overall estimate of system performance is possible. The effectiveness of insulation under different conditions can be observed by noting the change in condensate formed when the piping is exposed to variable conditions such as changing temperatures, humidity, rain and wind conditions.

For example, timed weighing of water coming from a heat exchanger indicates its heat transfer rate. Each pound of water is formed by the liberation of approximately 970 Btus in the steam system. (Often 1,000 Btus is used as a rule of thumb.)

This system has eight thousand feet of six inch steam piping. How much energy is being lost through the steam piping and condensate return piping?

1. The insulation heat loss averages 100 Btu per square foot per hour.

2. Pipe size 6 inches, 1.6 square foot per foot of length.

3. Condensate piping 1 3/4 inch average, 0.5 square foot per foot length.

Steam piping loss:
100 Btu/sq ft X 8,000 ft X 1.6 sq ft/ft = 1,280,000 Btu/Hr

Condensate piping loss:
100 Btu/sq ft X 8,000 ft X 0.5 sq ft/ft = 400,000 Btu/Hr

Total Loss: 1,680,000 Btu/Hr 6.7%

Condensate System Losses

Recovering hot condensate for reuse as boiler feed water is an important way to improve the efficiency of the system. The energy used to heat cold make-up water is a major part of the heat delivered for use by the steam system. For example heating a pound of cold make-up water requires an additional energy investment of 15% to 18% of the heat in the steam.

There are two problems to explore in this example. First 60% of the steam is not returning to the boiler plant as condensate and must be replaced with 42 F make up water. The second problem is the returning water is 160 F rather than a theoretical 212 F. How much is this costing in terms of energy loss?

(Note: A return temperature of 212 F is used for illustration, actually the condensate normally returns at a somewhat lower temperature.)

1. The condensate return temperature is 160 F with a loss of 52 Btu/lb. How much will this loss be with only a 40% condensate return?

17,900 LB/Hr X 52 Btu/lb X .4 = 372,320 Btu/hr

2. The make up water temperature is 42 F with a loss of 170 Btu/lb. How much will this loss be with 60% make up water being used?

17,900 Lb/Hr X 170 Btu/lb X .6 = 1,825,800 Btu/hr

Condensate recovery system energy loss:

Total Loss   2,198,120 Btu/hr 8.8%

Trap Flash Losses

When the hot condensate passes through a trap from the high pressure to the low pressure side, the water flashes off a percentage of steam because at the lower pressure the water can hold less energy. This energy difference is converted to flash steam. This steam which forms on the outlet side of the trap, like steam from the boiler contains latent heat that can be put back to work. Uses for this flash steam are heating coils, makeup water heating and other process water heating applications.

In the 250 PSI steam system, the steam and water temperature entering each steam trap is 401 F. As the condensate passes through the trap to atmospheric pressure it must give up some heat. This excess heat forms steam because water at 250 PSI and 401 F contains more energy that water at atmospheric pressure and 212 F. In this case the flash rate is 21%.

In an atmospheric return system, the flash steam can vent to atmosphere and be lost, how much energy is involved?

1. The steam flashed at atmospheric pressure contains 970 Btu/lb. (See steam tables in **Chapter 4**)

2   The condensate at 250 PSI contains 376 Btu/lb, at atmospheric pressure this value is 170 Btu/lb allowing for the heat in the make-up water of 10 Btu/lb at a temperature of 42 F.

Flash steam = (376 Btu/lb - 170 Btu/lb)/ 970 Btu/lb = 21%

Flash Loss = 17,900 LB/Hr X 1140 Btu/lb X 20% =

4.78 Million Btu/hr   19.1%

This estimate does not include steam losses caused by steam leaks in piping or steam blowing through traps.

Also, actual steam trap flash steam losses may be less where pressure reducing valves and other system conditions lower the pressure at the trap.

Pressurized condensate recovery systems are also effective in reducing trap flash losses by reducing the pressure across the traps and by collecting the flash steam at some central point to recover its energy.

Trap Leakage

Assuming that this system has 200 steam traps, it would not be unusual to find that 40% had problems. For the sake of illustration, lets assume 10% of the traps leaking at the rate of 50 pounds of steam per hour. (Based on published information, this rate is conservative)

Trap leakage can be caused by:

1. The slow closing of traps.

2. Partial closing.

3. A total failure of the trap to close at all.

4. Leaking in the closed position.

All these conditions allow steam to blow into the condensate system which is vented to the atmosphere.

Loss = 200 traps X 10% failed X 50 LB/Hr X 970 Btu/lb

Loss = 970,000 Btu/hr   3.9%

External Steam Leaks

Almost every system has some steam leaks. In this case there are four 4 foot plumes which are losing about 50 pounds of steam per hour each for a total of 200 pounds per hour and an 8 foot plume losing about 500 pounds per hour. How much is this costing?

Loss = 700 LB/Hr X 970 Btu/lb = 679,000 Btu/hr

Loss = 679,000 Btu/hr   2.7%

Only the latent heat of steam at atmospheric pressure of 970 Btu/lb is used for the heat loss value as the earlier calculation of condensate losses takes into account the sensible heat loss of the makeup water to replace the steam.

Internal Steam Leaks

Valves in bypasses around steam traps are a prime source of internal system leakage. Leaks inside the steam service system are invisible and hidden in the flash steam making it very difficult to detect. This steam leakage appears to be part of the system steam load even though it is venting to atmosphere through the open condensate recovery system.

In this case there are several bypass leaks amounting to 500 pounds of steam per hour.

Loss = 500 LB/Hr x 970 Btu/lb

Loss = 485,000 Btu/hr   1.9%

Inactive Steam Piping

Inactive steam piping, although not doing any work, is experiencing the same losses as the rest of the system. In this example 2,000 FT of a 8,000 FT system is inactive. Twenty-five percent of the system losses are in this section which is out of service costing about $130,000 a year.

Distribution System Efficiency

Loss Summary

Boiler Losses	Percent Loss	Annual Btu Loss (Millions) Loss	Annual Dollar
Stack Losses	22.8%	49,932	$273,600
Blowdown Losses	4.6%	10,074	$55,200
Surface Losses	1.0%	2,190	$12,000
Total	28.4%	62,196	$340,800

Distribution System Losses			
Insulation Losses	6.7%	14,717	$80,640
Condensate System Loss	8.8%	19,272	$105,600
Trap Flash Loss	19.1%	41,873	$229,440
Trap Leakage Loss	3.9%	8,497	$46,560
External Steam Leaks	2.7%	5,948	$32,592
Internal Steam Leaks	1.9%	4,249	$23,280
Total	43.2%	94,555	$518,112

Combined Losses	*71.6%*	*156,751*	*$858,912*
System Efficiency	*28.4%*		

Distribution System Efficiency

Conclusions:

1. The combined as-found efficiency for the boiler and distribution system is 28.4%.

2. The wasted energy is worth $858,912 a year.

3. The cost of fuel is $6.00 per million Btus and the cost of steam at the point of use is $21.12.

4. With a boiler input of 25 million Btus per hour only 7.1 million Btus per hour is reaching a point of use for productive work.

5. The inactive piping section is wasting about $130,000 a year or 25% of the total system loss.

Chapter 4

Energy Management Basics
for
Boilers and Systems

Energy Balance and Losses

Understanding how energy works in systems is essential to an effective energy management program, so let's examine some basics.

The purpose behind managing an efficient plant is to get as much energy as possible from the fuel. How much would this be?

A good starting point for establishing the "Ideal Scene" for an efficient plant is knowing how much energy is being delivered to the system. The term used to express the energy delivered to a plant in fuel is Higher Heating Value or (HHV)(Table 4.1).

The Higher Heating Value

"The total heat obtained from the combustion of a specified amount of fuel which is at 60 Deg F when combustion starts, and the combustion products of which are cooled to 60 Deg F before the quantity of heat released is measured".

The HHV represents the total amount of heat energy released from the fuel when it is completely burned.

For practical reasons, flue gas is not cooled down to 60 F because acid will form which can cause severe corrosion problems. Stack temperature is kept above the acid formation temperature wasting energy.

Fossil fuel heating values are expressed in terms of BTU/LB in tables and for standard calculations.

One Btu (British Thermal Unit) is defined as:

"The quantity of heat required to raise the temperature of one pound of water one degree fahrenheit at, or near, its point of maximum density 39.1 Deg F"

INPUT-OUTPUT EFFICIENCY

The starting point for finding efficiency begins with accurate knowledge of the amount of energy "input" into the process. Two things must be known, how much fuel is being fired and how many BTU's are in each pound of fuel. The range of heating values for various fuels is given in **Table 4.1**.

Boiler efficiency is the percentage of the fuel's higher heating value which is converted to steam. This is sometimes called the Fuel-To-Steam Efficiency. **Figure 4.1** shows the basics of Output/Input efficiency.

	As Fired Heating Value	
Fuel	*Btu/Lb*	*Other Units*
Natural Gas	20,000 - 23,500	950-1,100 Btu/SCF
#1 Fuel Oil	19,670 - 19,860	134-135 KBtu/gal
#2 Fuel Oil	19,179 - 19,750	138-142 KBtu/gal
#4 Fuel Oil	18,280 - 19,400	142-151 KBtu/gal
#5 Fuel Oil	18,100 - 19,020	142-149 KBtu/gal
#6 Fuel Oil	17,410 - 18,990	143-156 KBtu/gal
Bituminous Coal	11,500 - 14,500	NA
Subituminous Coal	11,500 - 14,500	NA
Lignite	5,000 - 8,300	NA
Wood-moisture free	8,000 - 10,100	NA

Table 4.1, Higher Heating Value (HHV) of various fuels.

Figure 4.1 Input/Output efficiency.

The formula for Input-Output efficiency is:

Efficiency = $\dfrac{\textbf{BTU Output}}{\textbf{BTU Input}}$ x 100

Efficiencies are limited by the acid dew point and excess air. The acid dew point is the temperature at which acid begins to form in the boiler exhaust. This minimum temperature does not allow for recovery of all the heat from the HHV of the as-fired fuel. Also, because extra air is required to insure complete combustion of the fuel at the burner the extra volume of exhaust gasses carry away additional waste heat.

Why Steam is Used

Steam is very useful because it automatically flows to the point of use and does not need to be pumped. It moves through the piping system to a point of lower pressure where steam is condensing into water as it gives up its latent heat.

Steam Properties

The steam has three properties that can be put to work; pressure, temperature and volume. Pressure and volume combine to drive machinery and temperature is used in processes using heat.

In the conversion of water from its liquid phase to steam, its vapor phase, heat is added to initially increase the water temperature to the boiling point.

Water heated under constant pressure will rise in temperature until a certain temperature is reached when it begins to

28

evaporate into steam or boil. The temperature at which water boils is known as **saturation temperature** and depends on the pressure under which the water is heated. If pressure increases the saturation temperature rises, if pressure decreases the saturation temperature falls. Each pressure condition has a corresponding saturation temperature. The temperature of the steam and water remain constant at each pressure.

Sensible heat is a rise in the measurable temperature of the water before it boils. This is what thermometers indicate. This is approximately one Btu per pound of water for degree of temperature rise.

When the water gets to its boiling point a change occurs, the water begins to evaporate with no change in temperature. This phase change from liquid to vapor absorbs energy known as the **Latent Heat of Evaporation**.

Steam that is not fully vaporized is called **wet steam** and the weight of water droplets in the wet steam compared to the weight of the steam is known as **% moisture**. Also, when heat is taken away from saturated steam condensation takes place creating moisture. There is great difficulty in obtaining steam that is exactly dry and saturated.

Steam that is heated above the saturation temperature is called **superheated steam.** This steam is dry and does not contain water droplets and it can have any temperature above the saturation temperature. This is usually done by applying heat to steam after it has been removed from contact with the water.

This phenomena also occurs when steam pressure is reduced by a pressure reducing valve (PRV) in the steam distribution system.

Accounting for Heat in Steam and Water

The heat content of water and steam is expressed in British Thermal Units (BTU) per pound and is known as **Enthalpy** expressed by the symbol **"h"** in tables and formulas.

The **"Steam Tables"** contain the basic information on the energy properties of steam. The information in this book was obtained from works of Keenan and Keys "Thermodynamic Properties of Steam" published by John Wiley and Sons, Inc.

Enthalpy of Saturated Liquid (h_f) is the heat required to rise the temperature of one pound of water from 32 F to the saturation temperature. This property is sometimes known as the **Heat of the Liquid**.

Enthalpy of Evaporation (h_{fg}) is the amount of heat required to change one pound of water at the saturation temperature to dry saturated steam at the same temperature. This is also known as the **Latent Heat of Evaporation**.

Enthalpy of Saturated Vapor (h_g) is the heat required to change one pound of water at 32 F into dry saturated steam. It is the sum of the enthalpys of saturated water and evaporation. It is also known as the **Total Heat of Steam**.

Table 4.1 lists the HHV for fuels and **Table 4.2** shows the BTU values for steam and **Table 4.3** the values for water. This

Gage Pressure (psig)	Saturation or Boiling Temperature (Degrees F)	Specific Volume (Cu. Ft./Lb.)	Heat Content Above 32 Degrees F		
			Sensible Heat or Heat of Liquid (Btu/lb.)	Latent Heat or Heat of Evaporation (Btu/lb.)	Total Heat (Btu/lb.)
0	212.0	26.80	180.1	970.3	1150.4
1	215.5	25.13	183.6	968.1	1151.7
2	218.7	23.72	186.8	966.0	1152.8
3	221.7	22.47	189.8	964.1	1153.9
4	224.5	21.35	192.7	962.3	1155.0
5	227.3	20.34	195.5	960.5	1156.1
6	229.9	19.42	198.2	958.8	1157.0
7	232.4	18.58	200.7	957.2	1157.9
8	234.9	17.81	203.2	955.6	1158.8
9	237.2	17.11	205.6	954.1	1159.7
10	239.5	16.46	207.9	952.5	1160.4
11	241.7	15.86	210.1	951.1	1161.2
12	243.8	15.31	212.2	949.7	1161.9
13	245.9	14.79	214.3	948.3	1162.6
14	247.9	14.31	216.4	946.9	1163.2
15	249.8	13.86	218.3	945.6	1163.9
16	251.7	13.43	220.3	944.3	1164.6
17	253.6	13.03	222.2	943.0	1165.2
18	255.4	12.66	224.0	941.8	1165.8
19	257.1	12.31	225.7	940.6	1166.3
20	258.8	11.98	227.5	939.5	1167.0
21	260.5	11.67	229.2	938.3	1167.5
22	262.2	11.37	230.9	937.2	1168.1
23	263.8	11.08	232.5	936.1	1168.6
24	265.4	10.82	234.1	935.0	1169.1
25	266.9	10.56	235.6	934.0	1169.6
30	274.1	9.45	243.0	928.9	1171.9
35	280.7	8.56	249.8	924.2	1174.0
40	286.8	7.82	256.0	919.8	1175.8
45	292.4	7.20	261.8	915.7	1177.5
50	297.7	6.68	267.2	911.8	1179.0
55	302.7	6.23	272.4	908.1	1180.5
60	307.3	5.83	277.2	904.6	1181.8
65	311.8	5.49	281.8	901.3	1183.1
70	316.4	5.18	286.2	898.0	1184.2
75	320.1	4.91	290.4	894.8	1185.2
80	323.9	4.66	294.4	891.9	1186.3
85	327.6	4.44	298.2	899.0	1187.2
90	331.2	4.24	301.9	886.1	1188.0
95	334.6	4.06	305.5	883.3	1188.8
100	337.9	3.89	308.9	880.7	1189.6
105	341.1	3.74	312.3	878.1	1190.4
110	344.2	3.59	315.5	875.5	1191.0
115	347.2	3.46	318.7	873.0	1191.7
120	350.1	3.34	321.7	870.7	1192.4

Table 4.2, "Steam Tables", heat content of steam and water at various pressures.

Gage Pressure (psig)	Saturation or Boiling Temperature (Degrees F)	Specific Volume (Cu. Ft./Lb.)	Heat Content Above 32 Degrees F		
			Sensible Heat or Heat of Liquid (Btu/lb.)	Latent Heat or Heat of Evaporation (Btu/lb.)	Total Heat (Btu/lb.)
125	352.9	3.23	324.7	868.3	1193.0
130	355.6	3.12	327.6	865.9	1193.5
135	358.3	3.02	330.4	863.7	1194.1
140	360.9	2.93	333.1	861.5	1194.6
145	363.4	2.84	335.8	859.3	1195.1
150	365.9	2.76	338.4	857.2	1195.6
155	368.3	2.68	340.9	855.0	1195.9
160	370.6	2.61	343.4	853.0	1196.4
165	372.9	2.54	345.9	850.9	1196.8
170	375.2	2.47	348.3	848.9	1197.2
175	377.4	2.41	350.7	846.9	1197.6
180	379.5	2.35	353.0	845.0	1198.0
185	381.6	2.30	355.2	843.1	1198.3
190	383.7	2.24	357.4	841.2	1198.6
195	385.8	2.19	359.6	839.2	1198.8
200	387.8	2.13	361.9	837.4	1199.3
210	391.7	2.04	366.0	833.8	1199.9
220	395.5	1.95	370.1	830.3	1200.4
230	399.1	1.88	374.1	826.8	1200.9
240	402.7	1.81	377.8	823.4	1201.3
250	406.1	1.74	381.6	820.1	1201.7
260	409.4	1.68	385.2	816.9	1202.1
270	412.6	1.62	388.7	813.7	1202.4
280	415.7	1.56	392.1	810.5	1202.7
290	418.8	1.52	395.5	807.5	1202.9
300	421.8	1.47	398.7	804.5	1203.2
400	448.2	1.12	428.1	776.4	1204.6
500	470.0	0.90	452.9	751.3	1204.3
600	488.8	0.75	474.6	728.3	1202.9

Table 4.2 (continued), "STEAM TABLES"; heat content of water and steam at various pressures.

Water Temperature (Degrees F)	Saturation Pressure (Inches of Mercury Vacuum or (psig)	Specific Volume (Cu. ft./lb.)	Density (lb./cu. ft.)	Weight (lb./gal.)	Specific Heat (Btu/lb. – Degrees F – Hr.)	Specific Gravity
32	29.8	.01602	62.42	8.345	1.0093	1.001
40	29.7	.01602	62.42	8.345	1.0048	1.001
50	29.6	.01603	62.38	8.340	1.0015	1.000
60	29.5	.01604	62.34	8.334	.9995	1.000
70	29.3	.01606	62.27	8.325	.9982	.998
80	28.9	.01608	62.19	8.314	.9975	.997
90	28.6	.01610	62.11	8.303	.9971	.996
100	28.1	.01613	62.00	8.289	.9970	.994
110	27.4	.01617	61.84	8.267	.9971	.991
120	26.6	.01620	61.73	8.253	.9974	.990
130	25.5	.01625	61.54	8.227	.9978	.987
140	24.1	.01629	61.39	8.207	.9984	.984
150	22.4	.01634	61.20	8.182	.9990	.981
160	20.3	.01639	61.01	8.156	.9998	.978
170	17.8	.01645	60.79	8.127	1.0007	.975
180	14.7	.01651	60.57	8.098	1.0017	.971
190	10.9	.01657	60.35	8.068	1.0028	.968
200	6.5	.01663	60.13	8.039	1.0039	.964
210	1.2	.01670	59.88	8.005	1.0052	.960
212	0.0	.01672	59.81	7.996	1.0055	.959
220	2.5	.01677	59.63	7.972	1.0068	.956
240	10.3	.01692	59.10	7.901	1.0104	.947
260	20.7	.01709	58.51	7.822	1.0148	.938
280	34.5	.01726	57.94	7.746	1.020	.929
300	52.3	.01745	57.31	7.662	1.026	.919
350	119.9	.01799	55.59	7.432	1.044	.891
400	232.6	.01864	53.65	7.172	1.067	.860
450	407.9	.0194	51.55	6.892	1.095	.826
500	666.1	.0204	49.02	6.553	1.130	.786
550	1030.5	.0218	45.87	6.132	1.200	.735
600	1528.2	.0236	42.37	5.664	1.362	.679
700	3079.0	.0369	27.10	3.623		.434

Table 4.3, thermal properties of water.

information combined serves as a common basis for Input-Output energy calculations.

The steam tables also show the saturation temperature changes with pressure. At atmospheric pressure water boils at 212 Deg F and at 250 psig the boiling temperature goes up to 400 Deg F. The working temperature of the steam is raised by increasing pressure **(Table 4.4)**

Pressure Psig	Temperature °F
5	227
10	240
15	250
20	259
25	267
50	298
75	320
100	338
150	353
200	388
250	406
300	422
400	448
500	470
600	488

Table 4.4 Temperature change with boiler pressure.

Energy recovery in the condensate system

On the condensate recovery side of the steam system, there is an important rule of thumb to know about. Heating water for use in the boiler uses a significant percentage of the total energy in the steam, so it is important to get condensate back to the plant as hot as possible.

The water being fed to the boiler must be heated to the boiling point to drive off oxygen and carbon dioxide which can cause severe pitting and corrosion to the boiler and piping systems.

The hotter the returning condensate is, the higher the efficiency of the system.

The Rule of Thumb for feedwater heating:

Every 11 Deg F that has to be added to the boiler feed water reduces efficiency by 1%.

Starting with the cold water side of the system 50 Deg F, water is raised to approximately 220 Deg F in the feed water heater, depending on the type of system, for injection into the boiler which takes one BTU per pound per degree F. In this case 170 BTU's per pound of water. This amounts to 15% of the energy in the steam.

Unfortunately, there can be another reason for boiling hot water returning to the boiler room. It is a sure sign that steam traps are leaking steam.

Calculating fuel savings and loss based on efficiency change

There is an important difference between efficiency improvement and fuel savings. An efficiency increase from 80% to 81% is a 1% efficiency improvement. It is a proportional increase of 1% out of 80%

(1/80%) or 1.25. This represents a 1.25% fuel savings from the 1% efficiency improvement.

The actual fuel savings percentage is always higher then the efficiency increase. Similarly the fuel savings loss percentage is always greater than the corresponding efficiency decrease.

The formula for fuel savings or loss resulting from the change in efficiency is:

$$Savings = \frac{New\ Efficiency - Old\ Efficiency}{New\ Efficiency}$$

Calculating performance deficiency costs

The cost benefit of maintaining boiler efficiency at a high level is easily calculated with this formula:

$$S = W_f \ X \ \frac{E_n - E_I}{E_n} \ X \ C_f \ X \ Hr$$

S is the potential fuel savings per year

W_f is the fuel use rate in million BTU/HR

E_I is the ideal efficiency

E_n is the new or existing efficiency

C_f is the cost of fuel per million BTU

Hr is operating hours per year

Because boiler efficiency usually changes with load, the potential for fuel savings will change with the typical load on the boiler based on the ideal or reference efficiency at the load being considered. This formula can be used for estimating energy saving at the typical load under study.

Example: a boiler is firing at 100 million BTU/HR and its efficiency had dropped from the ideal of 83% to 78%. The cost of fuel is currently $6.00 per million BTUs and it fires at this load for 6,000 hours a year.

This loss of efficiency will cost an estimated $230,400 in wasted fuel for the year if this loss of efficiency remains uncorrected.

It is useful to have a quick way to estimate potential fuel savings based on improvement in either net stack temperature or excess-air. **Figures 4.2 and 4.3** can be used as a quick reference for estimating savings potential.

34

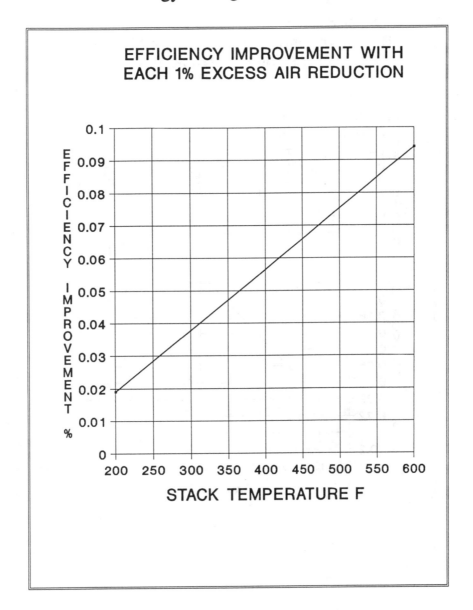

EFFICIENCY IMPROVEMENT WITH EACH 1% EXCESS AIR REDUCTION

Figure 4.2, Efficiency change (%) for each 1% change in excess air at various stack temperatures.

The efficiency improvement for each one percent change in excess air varies with the stack temperature **(Fig 4.2)**. To estimate efficiency change, multiply the factor (left) corresponding to the stack temperature times the change in excess air.

Example, what will the efficiency change be with a stack temperature of 500°F if the excess air is reduced by 50%? The factor from **Fig 4.2** at 500°F is .075; this multiplied by 50% excess air reduction is a 3.75% improvement in efficiency.

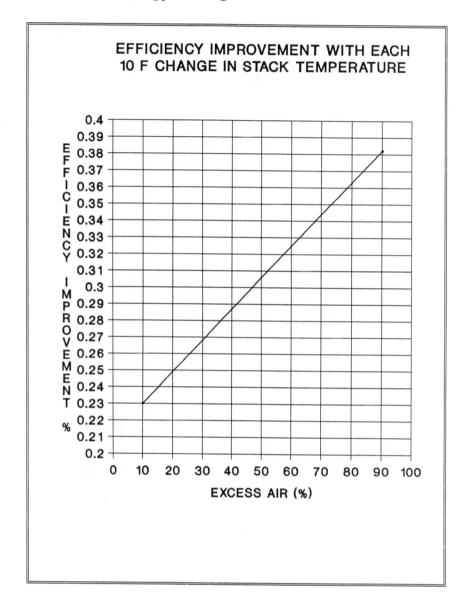

Figure 4.3, Efficiency change for each 10°F change in stack temperature for various levels of excess air.

The efficiency improvement for each 10 degrees F change in stack temperature varies with the excess air level **(Fig 4.3)**. To estimate efficiency change, multiply the factor (left) corresponding to the excess air level times each 10°F change in stack temperature.

Example, what will the efficiency change be with an excess air level of 60% if the stack temperature is reduced by 100°F? The factor from **Fig 4.3** at 60% excess air is 0.325; this multiplied by 100°F/10 a 3.25% improvement in efficiency.

Energy Management Basics

Improving Radiant Heat Transfer

The radiation section or that section of the boiler exposed to direct radiation from combustion, is the most effective heat transfer zone of a boiler. **Figure 4.4** shows that 8% of the heat exchange surface of a boiler is absorbing about 48% of the energy from combustion. The convection section, which comprises 34% of the heat exchange surface only absorbs about 20% of the heat of combustion.

Flame Temperature

Figure 4.5 shows that as the excess air is reduced, flame temperatures rise. Combustion flame temperatures are very complex to analyze and this chart should be treated only as an approximation. It can be seen in **Figure 4.5**, at 0% excess air, the flame temperature is about 3,200 F for natural gas and at 100% excess air this temperature drops to about 2,300 F. For number 6 oil these numbers are 4,200 F for 0% excess air and 2,400 F for 100% excess air.

The basic equation for radiant heat transfer, in the flame zone section, is:

$$Q = \rho ST^4$$

Q is BTU/HR

ρ is the Stefan-Boltzman constant, 1.17×10^{-9} Btu/sq ft, hr, T^4

S is surface area, sq ft

T is absolute temperature F + 460

It can be seen from the mathematics involved, the higher the temperature of the flame, the more intense the heat transfer. It becomes obvious that if the temperature of the radiant heat transfer section of the boiler is lowered by lower flame temperatures, there will be lower radiation heat transfer. This will ultimately increase flue gas exit temperature and decrease efficiency.

Stack temperatures have been observed to change by 50 F due to this phenomena. Changes in excess air directly affects flame temperature, and radiation heat transfer in the combustion zone.

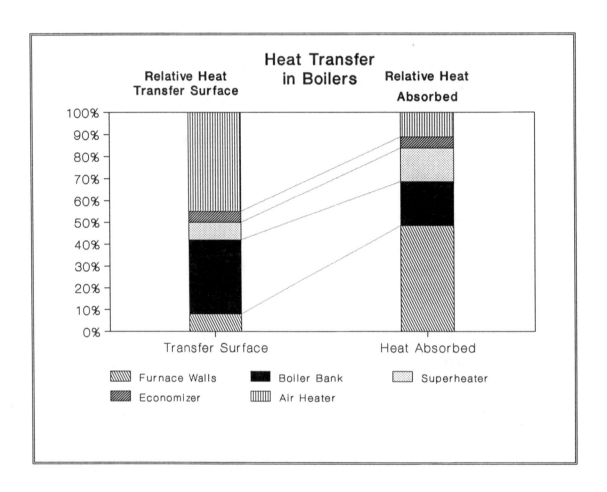

Figure 4.4, Graph of the ratio of relative heat transfer surface to heat absorbed in a boiler.

DETERMINING ACTUAL FLAME TEMPERATURE

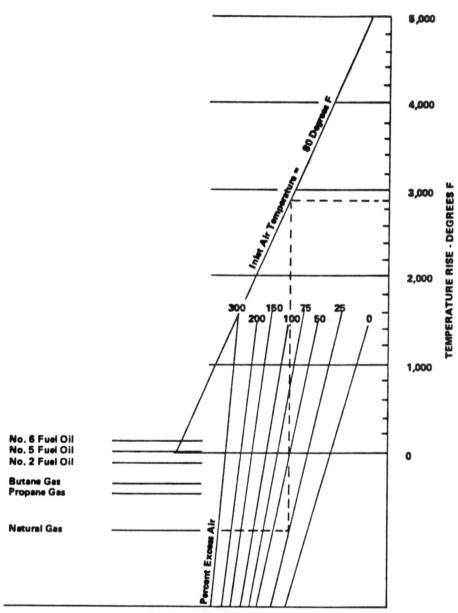

To determine the actual flame temperature in the combustion chamber, draw a horizontal line from the type of fuel being used to the point where it intersects the line corresponding to the percent excess air. From the point of intersection draw a vertical line upward to the point where it intersects the inlet air temperature line (80 degrees F). Then draw a horizontal line to the vertical axis -- Temperature Rise Degrees F.

Figure 4.5, Flame temperature variations with excess air.

Chapter 5

Efficiency Calculation Methods

A key part of a boiler efficiency improvement program is knowing the operating efficiency of the boiler and the corresponding increase in efficiency from as-found conditions to the final optimized condition. This may require several efficiency tests over an extended period .

The following paragraphs discuss the various test methods and computational procedures available for measuring efficiency.

ASME Computational Procedures

The basis for testing boilers is the American Society of Mechanical Engineers Power Test Code 4.1.

Figure 5.1 shows the ASME Test Form for Abbreviated Efficiency Test or so called "ASME Short form" which is used for both the Input-Output and heat Loss methods. **Figure 5.2** is the calculation sheet for the abbreviated ASME efficiency test.

Both Heat Loss and Input-Output boiler efficiency calculations are included in the ASME Short Form. This power test code has become the standard test procedure in many countries. It neglects minor efficiency losses and heat credits, considering only the Higher Heating Value of the input fuel.

Comparison of the Input-Output and Heat Loss Methods

Both methods are mathematically equivalent and would give identical results if the required heat balance (or heat loss) factors were considered in the corresponding boiler measurements could be performed without error.

When very accurate instrumentation and testing techniques are used, there is reasonably good agreement between the two calculation procedures. However, for practical boiler tests with limited instrumentation, comparisons between the two methods are generally poor. The poor results are primarily from the inaccuracies associated with the measurement of the flow and energy content of the input and output streams.

The efficiencies determined by these methods are "gross" efficiencies as opposed to "net" values which would include as additional heat input the energy required to operate all the boiler auxiliary equipment (i.e. combustion air fans, fuel pumps, fuel heaters, stoker drives, etc.)

These "gross" efficiencies can be considered essentially as the effectiveness of the boiler in extracting the available heat energy of the fuel.

It is important to take complete data when using this test form to fully document the test results no matter what procedure is used.

The Input-Output Method

Efficiency (percent) =

$$\frac{\text{Output}}{\text{Input}} \times 100$$

				TEST NO.		BOILER NO.	DATE
OWNER OF PLANT				LOCATION			
TEST CONDUCTED BY				OBJECTIVE OF TEST			DURATION
BOILER MAKE & TYPE						RATED CAPACITY	
STOKER TYPE & SIZE							
PULVERIZER, TYPE & SIZE						BURNER, TYPE & SIZE	
FUEL USED		MINE		COUNTY		STATE	SIZE AS FIRED

	PRESSURES & TEMPERATURES				FUEL DATA				
1	STEAM PRESSURE IN BOILER DRUM	psia			COAL AS FIRED PROX. ANALYSIS	% wt		OIL	
2	STEAM PRESSURE AT S. H. OUTLET	psia		37	MOISTURE		51	FLASH POINT F*	
3	STEAM PRESSURE AT R. H. INLET	psia		38	VOL MATTER		52	Sp. Gravity Deg. API*	
4	STEAM PRESSURE AT R. H. OUTLET	psia		39	FIXED CARBON		53	VISCOSITY AT SSU* BURNER SSF	
5	STEAM TEMPERATURE AT S. H. OUTLET	F		40	ASH		44	TOTAL HYDROGEN % wt	
6	STEAM TEMPERATURE AT R H INLET	F			TOTAL		41	Btu per lb	
7	STEAM TEMPERATURE AT R. H. OUTLET	F		41	Btu per lb AS FIRED				
8	WATER TEMP. ENTERING (ECON.) (BOILER)	F		42	ASH SOFT TEMP.* ASTM METHOD			GAS	% VOL
9	STEAM QUALITY % MOISTURE OR P.P.M.				COAL OR OIL AS FIRED ULTIMATE ANALYSIS		54	CO	
10	AIR TEMP. AROUND BOILER (AMBIENT)	F		43	CARBON		55	CH₄ METHANE	
11	TEMP AIR FOR COMBUSTION (This is Reference Temperature) †	F		44	HYDROGEN		56	C₂H₂ ACETYLENE	
12	TEMPERATURE OF FUEL	F		45	OXYGEN		57	C₂H₄ ETHYLENE	
13	GAS TEMP. LEAVING (Boiler) (Econ.) (Air Htr.)	F		46	NITROGEN		58	C₂H₆ ETHANE	
14	GAS TEMP. ENTERING AH (If conditions to be corrected to guarantee)	F		47	SULPHUR		59	H₂S	
	UNIT QUANTITIES			40	ASH		60	CO₂	
15	ENTHALPY OF SAT. LIQUID (TOTAL HEAT)	Btu/lb		37	MOISTURE		61	H₂ HYDROGEN	
16	ENTHALPY OF (SATURATED) (SUPERHEATED) STM.	Btu/lb			TOTAL			TOTAL	
17	ENTHALPY OF SAT. FEED TO (BOILER) (ECON.)	Btu/lb			COAL PULVERIZATION			TOTAL HYDROGEN % wt	
18	ENTHALPY OF REHEATED STEAM R.H. INLET	Btu/lb		48	GRINDABILITY INDEX*		62	DENSITY 68 F ATM. PRESS.	
19	ENTHALPY OF REHEATED STEAM R. H. OUTLET	Btu/lb		49	FINENESS % THRU 50 M*		63	Btu PER CU FT	
20	HEAT ABS/LB OF STEAM (ITEM 16 − ITEM 17)	Btu/lb		50	FINENESS % THRU 200 M*		41	Btu PER LB	
21	HEAT ABS./LB R.H. STEAM (ITEM 19 − ITEM 18)	Btu/lb		64	INPUT-OUTPUT EFFICIENCY OF UNIT %			ITEM 31 × 100 / ITEM 29	

							Btu/lb A. F. FUEL	% of A. F. FUEL
22	DRY REFUSE (ASH PIT + FLY ASH) PER LB AS FIRED FUEL	lb/lb			HEAT LOSS EFFICIENCY			
23	Btu PER LB IN REFUSE (WEIGHTED AVERAGE)	Btu/lb		65	HEAT LOSS DUE TO DRY GAS			
24	CARBON BURNED PER LB AS FIRED FUEL	lb/lb		66	HEAT LOSS DUE TO MOISTURE IN FUEL			
25	DRY GAS PER LB AS FIRED FUEL BURNED	lb/lb		67	HEAT LOSS DUE TO H₂O FROM COMB. OF H₂			
	HOURLY QUANTITIES			68	HEAT LOSS DUE TO COMBUST. IN REFUSE			
26	ACTUAL WATER EVAPORATED	lb/hr		69	HEAT LOSS DUE TO RADIATION			
27	REHEAT STEAM FLOW	lb/hr		70	UNMEASURED LOSSES			
28	RATE OF FUEL FIRING (AS FIRED wt)	lb/hr		71	TOTAL			
29	TOTAL HEAT INPUT (Item 28 × Item 41)/1000	kB/hr		72	EFFICIENCY = (100 − Item 71)			
30	HEAT OUTPUT IN BLOW-DOWN WATER	kB/hr						
31	TOTAL HEAT OUTPUT (Item 26×Item 20)+(Item 27×Item 21)+Item 30 / 1000	kB/hr						

	FLUE GAS ANAL. (BOILER) (ECON) (AIR HTR) OUTLET		
32	CO₂	% VOL	
33	O₂	% VOL	
34	CO	% VOL	
35	N₂ (BY DIFFERENCE)	% VOL	
36	EXCESS AIR	%	

* Not Required for Efficiency Testing

† For Point of Measurement See Par. 7.2.8.1-PTC 4.1-1964

Figure 5.1, ASME test form summary sheet for abbreviated efficiency testing.

CALCULATION SHEET   **ASME TEST FORM**
FOR ABBREVIATED EFFICIENCY TEST

	OWNER OF PLANT	TEST NO.	BOILER NO.	DATE

30 — HEAT OUTPUT IN BOILER BLOW-DOWN WATER = LB OF WATER BLOW-DOWN PER HR $\times \left[\dfrac{\text{ITEM 15} - \text{ITEM 17}}{1000}\right]$ = ... kB/hr

24 —

If impractical to weigh refuse, this item can be estimated as follows

DRY REFUSE PER LB OF AS FIRED FUEL $= \dfrac{\%\ \text{ASH IN AS FIRED COAL}}{100 - \%\ \text{COMB. IN REFUSE SAMPLE}}$

CARBON BURNED PER LB AS FIRED FUEL $= \dfrac{\text{ITEM 43}}{100} - \left[\dfrac{\text{ITEM 22} \times \text{ITEM 23}}{14,500}\right]$ = ...

NOTE: IF FLUE DUST & ASH PIT REFUSE DIFFER MATERIALLY IN COMBUSTIBLE CONTENT, THEY SHOULD BE ESTIMATED SEPARATELY. SEE SECTION 7, COMPUTATIONS.

25 —

DRY GAS PER LB AS FIRED FUEL BURNED $= \dfrac{11CO_2 + 8O_2 + 7(N_2 + CO)}{3(CO_2 + CO)} \times \left(\text{LB CARBON BURNED PER LB AS FIRED FUEL} + \dfrac{3}{8}S\right)$

$= \dfrac{11 \times \text{ITEM 32} + 8 \times \text{ITEM 33} + 7\left(\text{ITEM 35} + \text{ITEM 34}\right)}{3 \times \left(\text{ITEM 32} + \text{ITEM 34}\right)} \times \left[\text{ITEM 24} + \dfrac{\text{ITEM 47}}{267}\right]$ = ...

36 —

EXCESS AIR† $= 100 \times \dfrac{O_2 - \dfrac{CO}{2}}{.2682 N_2 - \left(O_2 - \dfrac{CO}{2}\right)} = 100 \times \dfrac{\text{ITEM 33} - \dfrac{\text{ITEM 34}}{2}}{.2682(\text{ITEM 35}) - \left(\text{ITEM 33} - \dfrac{\text{ITEM 34}}{2}\right)}$ = ...

HEAT LOSS EFFICIENCY	Btu/lb AS FIRED FUEL	$\dfrac{\text{LOSS}}{\text{HHV}} \times 100 =$	LOSS %
65 HEAT LOSS DUE TO DRY GAS = $\dfrac{\text{LB DRY GAS}}{\text{PER LB AS FIRED FUEL}} \times C_p \times (t_{lvg} - t_{air}) = \text{ITEM 25} \times 0.24 \dfrac{(\text{ITEM 13}) - (\text{ITEM 11})}{\text{Unit}}$ = ...	. . . . . .	$\dfrac{65}{41} \times 100 =$	. . . . . .
66 HEAT LOSS DUE TO MOISTURE IN FUEL = $\dfrac{\text{LB }H_2O\text{ PER LB}}{\text{AS FIRED FUEL}} \times \big[(\text{ENTHALPY OF VAPOR AT 1 PSIA \& T GAS LVG}) - (\text{ENTHALPY OF LIQUID AT T AIR})\big] = \dfrac{\text{ITEM 37}}{100} \times \big[(\text{ENTHALPY OF VAPOR AT 1 PSIA \& T ITEM 13}) - (\text{ENTHALPY OF LIQUID AT T ITEM 11})\big]$ = ...	. . . . . .	$\dfrac{66}{41} \times 100 =$	. . . . . .
67 HEAT LOSS DUE TO H_2O FROM COMB. OF H_2 = $9H_2 \times \big[(\text{ENTHALPY OF VAPOR AT 1 PSIA \& T GAS LVG}) - (\text{ENTHALPY OF LIQUID AT T AIR})\big]$ $= 9 \times \dfrac{\text{ITEM 44}}{100} \times \big[(\text{ENTHALPY OF VAPOR AT 1 PSIA \& T ITEM 13}) - (\text{ENTHALPY OF LIQUID AT T ITEM 11})\big]$ = ...	. . . . . .	$\dfrac{67}{41} \times 100 =$	. . . . . .
68 HEAT LOSS DUE TO COMBUSTIBLE IN REFUSE = $\text{ITEM 22} \times \text{ITEM 23}$ = ...	. . . . . .	$\dfrac{68}{41} \times 100 =$	. . . . . .
69 HEAT LOSS DUE TO RADIATION* $= \dfrac{\text{TOTAL BTU RADIATION LOSS PER HR}}{\text{LB AS FIRED FUEL} - \text{ITEM 28}}$	. . . . . .	$\dfrac{69}{41} \times 100 =$	. . . . . .
70 UNMEASURED LOSSES **	. . . . . .	$\dfrac{70}{41} \times 100 =$	
71 TOTAL	. . . . . .	. . . . . .	. . . . . .
72 EFFICIENCY = (100 − ITEM 71)	. . . . . .	. . . . . .	. . . . . .

† For rigorous determination of excess air see Appendix 9.2 – PTC 4.1-1964
* If losses are not measured, use ABMA Standard Radiation Loss Chart, Fig. 8, PTC 4.1-1964
** Unmeasured losses listed in PTC 4.1 but not tabulated above may by provided for by assigning a mutually agreed upon value for Item 70.

Figure 5.2, ASME calculation sheet for abbreviated efficiency testing.

This method measures the heat absorbed by the water and steam and compares it to the total energy input of the higher heating value (HHV) of the fuel.

This method requires the accurate measurement of fuel input. Also, accurate data must be available on steam pressure, temperature and flow, feed water temperature, stack temperature, and air temperature to complete energy balance calculations.

Figure 5.3 illustrates the envelope of equipment included in the designation of "Steam Generating Unit."

Figure 5.4 shows the relationship between input, output, credits and losses.

Because of the many physical measurements required at the boiler and the potential for significant measurement errors, the Input-Output method is not practical for field measurements at the majority of industrial and commercial boiler installations where precision instrumentation is not available.

Large errors are possible because this method relies on the difference in large numbers. If the steam flow is off by 2-3% and other instrumentation have a similar level of error, then the cumulative error can become unacceptable, producing false information.

The Input-Output test method is also very labor intensive. Precision instrumentation must be specified and installed. Test runs are usually more than four hours and must be rejected for any inconsistent data. Trial runs are often required to check out instrumentation and identify problems with the boiler as well as to train test personnel and observers.

Often plants cannot support testing for long periods because of load considerations. For example, a plant may not be able to provide either full load or partial load conditions for extended periods for various reasons which can cause premature curtailment of tests.

Problems like inconsistent water level control or variations in outlet steam pressure can prevent the stable thermal balance required for accurate test information.

The Heat Loss Method

Efficiency (%) = 100 % - Heat Loss %

The Heat Loss method subtracts individual energy losses from 100% to obtain percent efficiency. It is recognized as the standard approach for routine efficiency testing, especially at industrial boiler sites where instrumentation quite often is minimal.

The losses measured are:

1. Heat loss due to dry gas.

2. Heat loss due to moisture in fuel.

3. Heat loss due to H_2O from the combustion of hydrogen.

4. Heat loss due to combustibles in refuse. (for coal)

5. Heat loss due to radiation.

6. Unmeasured losses.

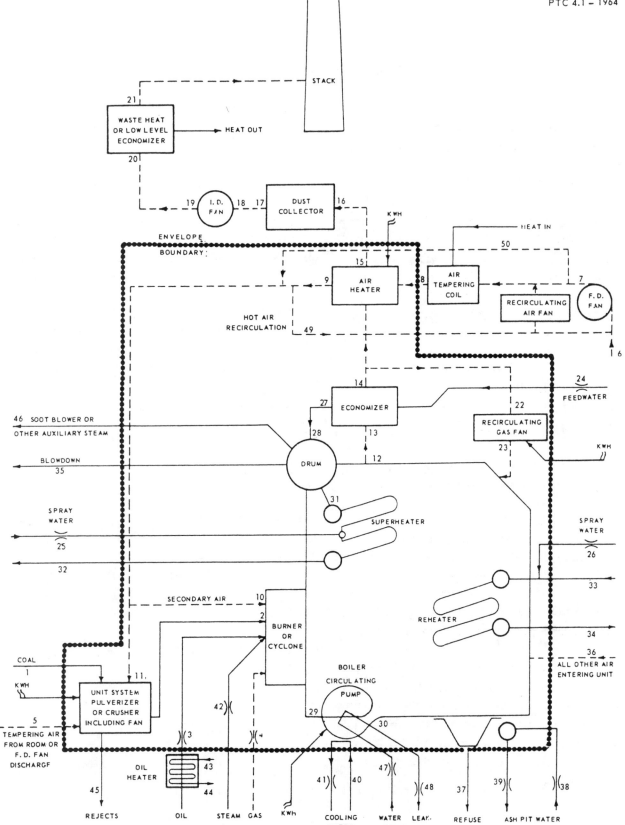

Figure 5.3, ASME designated envelop for steam generating unit.

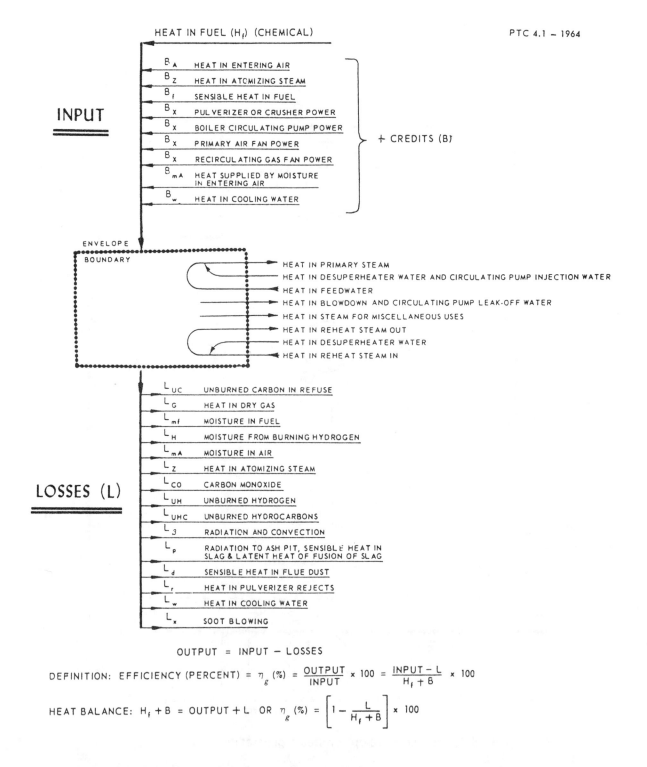

OUTPUT = INPUT − LOSSES

DEFINITION: EFFICIENCY (PERCENT) $= \eta_g \, (\%) = \dfrac{\text{OUTPUT}}{\text{INPUT}} \times 100 = \dfrac{\text{INPUT} - L}{H_f + B} \times 100$

HEAT BALANCE: $H_f + B = \text{OUTPUT} + L$ OR $\eta_g \, (\%) = \left[1 - \dfrac{L}{H_f + B} \right] \times 100$

Figure 5.4, ASME Heat Balance for steam generators.

46

This procedure neglects minor efficiency losses and heat credits and only considers the chemical heat (Higher Heating Value) of the fuel as energy input.

In addition to being more accurate for field testing, the heat loss method identifies exactly where the heat losses are thus aiding energy savings efforts.

This method might be termed the flue gas analysis approach since the major heat losses considered by this method are based on measured flue gas conditions at the boiler exit together with an analysis of the fuel composition.

This method requires the determination of the exit flue gas excess O_2 (or CO_2), CO, combustibles, temperature and the combustion air temperature.
The heat loss method is a much more accurate and more accepted method of determining boiler efficiencies in the field provided that the measurements of the flue gas conditions are accurate and not subject to air dilution or flue gas flow stratification or pocketing

Combustion Heat Loss Tables

Tables for stack gas heat losses for different types of fuels have been prepared separately and are available for determining flue gas losses due to dry gas hydrogen and moisture.

Heat Loss Due to Radiation

Radiation loss, not associated with the flue gas conditions can be estimated from the standard curve given in **Figure 5.5**.

Radiation loss can also be measured using a simple direct method, using an infrared instrument which has specially designed to detect radiation losses from the boiler surface and gives a read out in BTU/SQFT/HR.

The technique is to make a drawing of the boiler surface dimensions and make grids of equal areas with the average measured heat loss for each grid then adding up the losses for all the surface grids.

Comparing Methods for Measuring Boiler Efficiency

Input-Output Method

    a. Most direct method

    b. Difficult and expensive to measure accurately

    c. Does not locate energy losses

Heat Loss Method

    a. Indirect method (100% - energy losses)

    b. Simple and accurate

    c. locates and sets magnitude of energy losses and therefore is a key to efficiency improvement efforts.

    d. Allows assessment of potential efficiency improvements and energy savings.

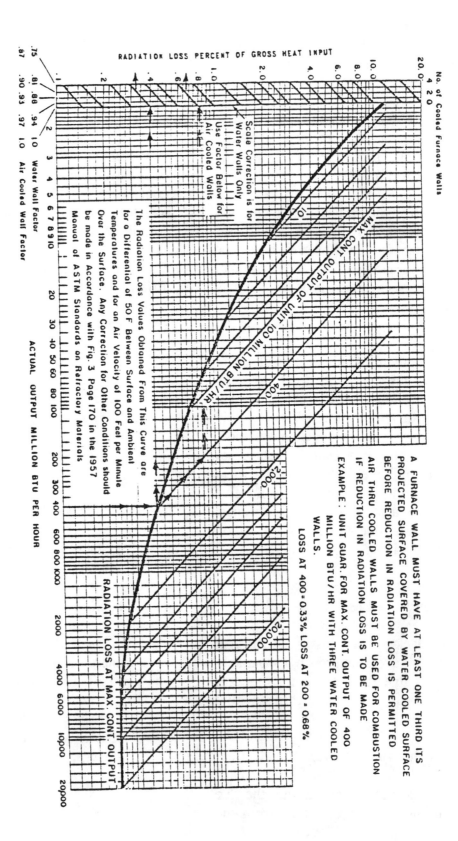

Figure 5.5, Standard Radiation Heat Loss Chart. (Courtesy of the American Boiler Manufacturers Association.)

48

Efficiency Calculation Metnods

Stack Gas Heat Loss Solution

To calculate the stack heat loss using the ASME Short Form, it is necessary to make separate calculations of the heat losses due to dry gas, moisture in the fuel and moisture in the flue gas due to hydrogen in the fuel. The three heat losses are then totaled to form the stack gas heat loss.

Calculation of Stack Gas Heat Loss

Heat loss due $= P_f \times 0.24 \times (T_g - T_a) =$ BTU/LB (As Fired Fuel)
to Dry Gas

$$P_f = \frac{LB\ Dry\ Gas}{LB\ as\ fired\ fuel} = \frac{11\ CO_2 + 8\ O_2 + 7(CO + N_2)}{3(CO2 + CO)} \times \left(\frac{\%C}{100} + \frac{\%S}{267} \right)$$

T_g = Flue gas temperature leaving unit, degrees F

T_a = Combustion air temperature entering unit, degrees F

%C, %S = Percent by weight in fuel analysis of carbon and sulfur

O_2, CO_2, CO, N_2 = Percent by volume in flue gas.

Heat Loss Due to Moisture $= 9 \frac{(H_2)}{100} \times (h_g - h_a) =$ BTU/LB (As Fired Fuel)
formed from H_2 in fuel

Heat Loss Due to $=$ % Moisture $\times (h_g - h_a) =$ BTU/LB (As Fired Fuel)
Moisture in Fuel

H_2 = % hydrogen in as fired fuel by weight

% Moisture = % moisture in as fired fuel by weight

h_g = Enthalpy of water vapor at 1 psia and T_g

h_a = Enthalpy of liquid water at T_a

Excess Air Solution

$$Excess\ Air = 100 \times \frac{O_2 - CO/2}{.2682N_2 - (O_2 - CO/2)}$$

49

Chapter 6

Combustion Analysis

COMBUSTION ANALYSIS

Understanding the combustion process is very important for safe and efficient operation of a plant. Perfect combustion is the proper mixture of fuel and air under exacting conditions where both the oxygen and the fuel are completely consumed in the combustion process. Having just the right amount of oxygen (no more, no less) is called the **stoichiometric** point, simply the ideal air to fuel ratio for combustion.

Anything above the ideal amount of air supplied to the combustion process is called **Excess Air**, and is wasteful.

A CO_2 analysis alone does not provide a safe indication of the combustion air/fuel setting. Additional requirements of either smoke or CO is recommended as the same CO_2 measurements can occur on either side of stoichiometric.

Excess air is the preferred term to describe the combustion setting on the safe side of stoichiometric. Using oxygen measurements is the best way find excess air.

By understanding a few simple instruments most of the potentially hazardous conditions can be reduced. The key is to properly measure smoke, oxygen, carbon monoxide, draft and gas pressure.

Caution

A most dangerous approach, when dealing with combustion systems is thinking the systems will always be correct and not considering that it can be affected by small and seemingly unrelated external forces.

The following is a list of key parameters considered as important safety measurement areas:

- Input gas pressure
- Draft (overfire and boiler exhaust)
- Carbon Monoxide
- Stack temperature
- Smoke
- Combustibles

Too much smoke is one of the most common indicators of excessive fuel wastage and can cause major problems.

Stack Fires:

Build up of combustibles in the exhaust system and chimneys can cause fires and explosions. To a lesser extent a build-up of soot in the exhaust system can block the normal passage of flue gases further restricting the amount of oxygen supplied for combustion progressively compounding the problem.

Sources of Problems

Four basic combustion zone conditions that prevent clean, efficient combustion are:

- Insufficient combustion air applied to the flame to permit clean combustion at an acceptable combustion efficiency

- Non-uniform delivery of fuel/or

combustion air to the combustion zone.

- Insufficient temperature of the combustion zone to permit proper burning of the fuel.

- Insufficient flame turbulence or inadequate mixing of fuel and air during the vaporization and burning.

These points are precisely why the smoke test is an indispensable part of oil burner servicing, accurate smoke testing takes less than a minute.

The Smoke Spot Test

The smoke test method has long been recognized as the acceptable standard for oil burners. It is used by Underwriters Laboratories in approval testing of oil burners and it is the test method specified by the National Oil Fuel Institute and by the U.S. Department of Commerce, in their standard for testing materials; (D2156-65) and also in the European standard DIN 51 402.

All three components of the smoke test set (pump, scale, and filter paper) must meet the U.S. commercial standards **(Figure 6.1)**.

The ten spots on the scale range in equal photometric steps from white to black to completely cover any smoke condition which may be experienced.

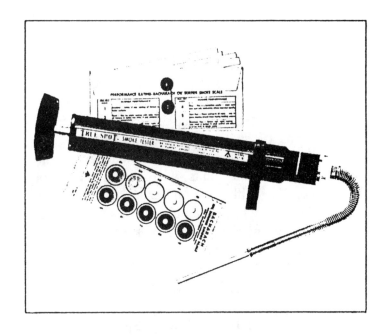

Figure 6.1, Bacharach smoke test kit.

Use of the smoke tester, three simple steps (Figure 6.2).

1. Clamp the filter strip into the test pump and insert into the flue.

2. pump handle through ten even unhurried strokes.

2. Remove filter paper strip and place between smoke scale and white reflective and compare smoke test spot on filter paper with smoke spots on scale. The smoke reading is the closest match.

Combustion Analysis

Interpreting the results

Not all kinds of oil burners will be equally affected by the same smoke content in the flue gas, this fact is shown in **Figure 6.3** which interprets smoke scale readings in terms of sooting produced. Depending on the construction of the heat exchanger or the boiler, some units will accumulate soot rapidly with a number 3 smoke spot number, while accumulation of soot on other units at the same smoke spot reading may be relatively slower.

Possible causes of smoke conditions:

Condition: If the Smoke Spot Number is too high, soot deposits in furnace and on heat exchange surfaces will lead to poor efficiency.

Cause:

1. If excess air is low 5%-20%

    a. Overfiring

    b. Too little excess air

2. If excess air is high above 50%

    a. Faulty nozzle and inefficient atomization of fuel.

    b. Combustion chamber trouble

    c. Chilling of the combustion process before complete combustion occurs.

High Stack Temperature

A high stack temperature may indicate any of the following conditions and should be immediately checked and remedied:

1. Soot deposits on heat exchange surfaces.

2. Short circuits of hot combustion gases due to problems with baffles.

3. Overfiring, check fuel rate.

4. Water side scale deposits from improper water treatment.

5. High excess air which reduces combustion chamber heat transfer.

Low Stack Temperature

If the stack temperature drops lower than normal, lower than 250 F for natural gas or 275-300 F for oil.

1. Possible underfiring.

2. Incomplete combustion of gas with dangerous carbon monoxide production.

3. Possible ruptured boiler tube or other component which is cooling the gases with steam or water.

4. In negative draft units, cold air may be entering boiler through open door or defective wall or skin.

The Dangers of Carbon Monoxide

An example of the dangers of carbon monoxide is evident in a report from Canada. Between 1973 and 1983 there were 293 reports of carbon monoxide poisoning,

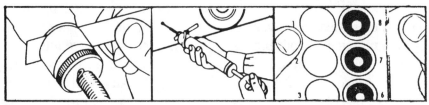

Figure 6.2, Smoke spot testing procedure.

Effect of Smoke on Burner Performance		
Smoke Scale Number	Rating	Sooting Produced
1	Excellent	Extremely light if at all
2	Good	Slight sooting that may not increase stack temperature appreciably
3	Fair	May be some sooting but will rarely require cleaning more than once a year
4	Poor	Borderline condition. Some units will require cleaning more than once a year
5	Very Poor	Sooting occurs rapidly and heavily

Figure 6.3, Smoke spot rating scale.

including 145 deaths. Also, combustion systems caused 238 deaths during the same period.

What is carbon monoxide?

It is the product of incomplete combustion and is a flammable colorless and odorless gas. Carbon monoxide is about the same density as the air that we breath. Therefore, it easily mixes to form a deadly atmosphere.

The major hazards of carbon monoxide are its toxicity and flammability. Carbon monoxide becomes a combustible gas when its concentration reaches 12.5% by volume (125,000 ppm).

Carbon monoxide is classified, however, as a chemical asphyxiant which produces a toxic action by preventing the blood from absorbing oxygen. Since the affinity of carbon monoxide is 200-300 times that of oxygen in blood, even small amounts of carbon monoxide in the air will cause toxic reactions to occur.

If breathed for a sufficiently long time, a carbon monoxide concentration of only 50 ppm will produce symptoms of poisoning. As little as 200 ppm will produce slight symptoms like a headache or discomfort in just a few hours. A concentration of 400 ppm will produce a headache and discomfort in two to three hours. The effect at higher concentrations may be so sudden that a person has little or no warning before collapsing. It should be noted that all of these values are approximate and vary as to the individual.

To prevent over-firing and under-firing the fuel supply pressure must be tested from time to time to insure the firing rate has not shifted because of a pressure change.

In small boilers or furnaces, a draft measurement is necessary to guard against a gas reversal where exhaust gases are escaping to the environment, which is potentially toxic. A small draft gage will indicate problems, like building exhaust fans pulling gasses back down the chimney, obstructions in the exhaust system, down drafts from high wind conditions and defective stack covers and other problems.

Chimney Effect

When taking draft readings insure that the system is warmed up to normal operating temperatures. As air is warmed, it expands and the same weight of air will take up more space becoming lighter. This warm "light" air will rise up the chimney decreasing the furnace pressure. If this low pressure is not established, combustion products may escape. If the draft is too high, and the hot gasses are creating too much negative draft heat will be lost up the stack.

Condensing Flue Gases to Improve Efficiency

Fuel is a hydrocarbon which means that it is made up of hydrogen and carbon. Carbon burns dry but each pound of hydrogen that enters into the combustion process forms about 9 pounds of water. Now, at the 2,000 to 3,000 degree combustion temperature this water is in the form of steam and it carries a considerable amount of latent heat. If this latent heat can be extracted from the exhaust gasses, there is an opportunity to raise efficiency by 10% or more.

Roughly there will be about 970 BTUs available from each pound of water that is condensed in the flue gas. In flue gas condensing systems the exit temperature is typically about 100 F.

Sulfur in Fuel Forms Acid

Sulfur in some fuels can end up as sulfuric acid when the flue gas temperatures drop too low. Boiler damage and corrosion from sulfuric acid has been a problem and a challenge for many years, causing large (energy wasting) safety margins in stack temperature to be used to avoid damage.

In the past temperatures were maintained above the approximate levels listed below to prevent formation of SO_2 and SO_3 which combines with moisture to form acids.

- Natural Gas   250 F
- No. 2 Heating Oil   275 F
- No. 6 Fuel Oil   300 F
- Coal   325 F
- Wood   400 F

Combustibles

Because combustibles in the flue gas are unburned fuels, this represents fuel flowing out of the stack.

Scientists have observed on occasion that combustibles are composed of equal parts carbon monoxide and hydrogen. Hydrogen has a heating value of 61,100 BTU/Lb, Carbon Monoxide has a heating value of 4,347 BTU/Lb.

Combustion Efficiency

In practice combustion efficiency is thought of as the total energy contained per pound of fuel minus the energy carried away by the hot flue gasses exiting through the stack, expressed as a percentage.

Combustion efficiency is only part of the total efficiency. Radiation loss from hot exposed boiler surfaces, blowdown losses and electrical losses in pumps and fans are examples of other kinds of losses that must be considered in determining total efficiency. However in most fuel burning equipment, the most effective way to reduce wasted fuel is to improve combustion efficiency. To do so, it is necessary to understand the fundamentals of combustion.

Stoichiometric Combustion

The three essential components of combustion are fuel, air and heat. In fossil fuels, there are really only three elements of interest: carbon, hydrogen and sulfur.
During combustion, each reacts with oxygen to release heat:

$$C + O_2 ---> CO_2 + 14,093 \text{ Btu/lb}$$

$$H_2 + {}_{1/2}O_2 ----> H_2O + 61,100 \text{ Btu/lb}$$

$$S + O_2 ----> SO_2 + 3,983 \text{ Btu/lb}$$

Pure carbon, hydrogen and, sulfur are rarely used as fuels. Instead, common fuels are made up of chemical compounds containing these elements. Methane, for example, is a hydrocarbon gas that burns as follows:

$$CH_4 + 2O_2 ----> \\ CO_2 + 2H_2O + 1,013 \text{ Btu/Ft}^3$$

Pure oxygen is also rarely used for combustion. Air contains about 21 percent oxygen and 79 percent nitrogen by volume and is much more readily available than pure oxygen:

$$CH_4 + 2O_2 + 7.53N_2 ---->$$
$$CO_2 + 2H_2O + 7.53N_2 + 1,013 \text{ Btu/Ft}^3$$

In this example, one cubic foot of methane (at standard temperature and pressure) will burn completely with 9.53 cubic feet of air containing 21 percent oxygen and 79 percent nitrogen. This complete burning of fuel, with nothing but carbon dioxide, water, and nitrogen as the end product is known as **stoichiometric** combustion **(Figure 6.4)**. The ratio of 9.53 cubic feet of air to one cubic foot of methane is known as the **stoichiometric** air/fuel ratio. The heat released when the fuel burns completely is known as the **heat of combustion**.

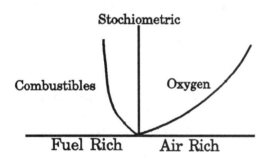

Figure 6.4. Stoichiometric point and air-rich and fuel-rich combustion shown as a function of the air fuel ratio.

The Importance of Excess Air

As most combustion equipment operators know, it is extremely undesirable to operate a burner with less-than-stoichiometric combustion air. Not only is this likely to result in smoking but it will significantly reduce the energy released by the fuel.

If a burner is operated with a deficiency of air, carbon monoxide and hydrogen will appear in the products of combustion. CO and H_2 are the result of incomplete combustion and are known as **combustibles.** Anything more than a few hundred parts per million of combustibles in flue gas indicates inefficient burner operation.

In actual applications, it is impossible to achieve stoichiometric combustion because burners can not mix fuel and air perfectly. To insure that all of the fuel is burned and little or no combustibles appear in the flue gas, it is common practice to supply some amount of excess air. In the era of cheap energy it was not uncommon to run a burner with a large amount of excess air in order to avoid smoking. Today this is becoming known for the highly wasteful practice it really is.

How do you achieve Optimum Combustion Efficiency?

Too little excess air is inefficient because it permits unburned fuel, in the form of combustibles, to escape up the stack. But too much excess air is also inefficient because it enters the burner at ambient temperature and leaves the stack hot, thus stealing useful heat from the process. This leads to the fundamental rule:

"Maximum combustion efficiency is achieved when the correct amount of excess air is supplied so the sum of both unburned fuel loss and flue gas heat loss is minimized".

Measuring Combustibles

Combustible analyzers are available to accurately measure CO and H_2 concentrations in flue gas to (+/-) 10 ppm or less. Carbon monoxide analyzers are often used in control systems because of their greater accuracy and calibration stability.

Flue Gas heat loss

Flue gas heat loss is the largest single energy loss in every combustion process. It is generally impossible to eliminate flue gas heat losses because the individual constituents of flue gas all enter the system cold and leave at elevated temperatures. Flue gas heat loss can be minimized by reducing the amount of excess air supplied to the burner.

Flue gas heat loss increases with both increasing excess air and temperatures. As both the carbon dioxide and oxygen level in flue gases are directly related to the amount of excess air supplied, either a CO_2 or an O_2 flue gas analyzer can be used to measure this loss. However, in recent years, CO_2 analysis has fallen out of favor.

There are a number of problems when CO_2 is used for analysis. CO_2 can be measured on both sides of the stoichiometric mix bringing about confusion about air deficiency or excess air (**Figure 6.5**). Also CO_2 readings may not be correct when different fuels having different hydrocarbon ratios are used.

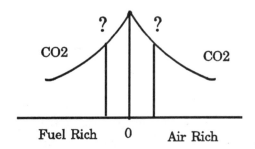

Figure 6.5, CO2 can be measured on both sides of stoichiometric. This can lead to questions about the actual air-fuel ratio being either air-rich or fuel-rich.

Both of these errors are unacceptable in modern combustion control systems. The development of improved oxygen analyzers has all but eliminated the use of carbon dioxide flue gas analyzers.

Flue gas measurement

Orsat testing

One of the earliest methods of measurement is still in use today. The orsat test is a manually-performed test in which a flue gas sample is passed successively through a series of chemical reagents. The chemicals each absorb a single gas constituent, usually carbon dioxide, oxygen, and carbon monoxide. After the sample passes through each reagent, its volume is accurately measured. The reduction in volume indicates the amount of gas that was originally in the sample.

There are several disadvantages to using the Orsat flue gas testing apparatus:

    a. It is slow, tedious work

b. Its accuracy is affected by the purity of the reagents

c. Operator skill is very demanding and under field conditions expert control is necessary to prevent data scatter and test rejection

d. Most important, the Orsat test measures only small samples and an unacceptable amount of time can go by before the unit is ready to analyze another sample, quite possibly missing information on the actual dynamics of the combustion systems operation

e. All data must be hand recorded. Computerized data systems are becoming increasingly important in combustion system analysis because of their continuous flow of data and automatic record keeping capabilities.

Measuring Carbon Dioxide (CO₂) and Oxygen (O₂) using chemical absorption instruments.

Because of the complexity and the operator skills demanded by the Orsat flue gas analyzer, simpler less complex devices have been developed using the chemical absorption process.

Notable among the test instruments now in daily use in many plants using this technique are instruments like the Bacharach Fyrite CO₂ and O₂ indicators (**Figure 6.6**).

The primary difference between the O₂ and CO₂ indicators are the chemicals used to absorb these gasses. They are a very practical, simple, rugged and economical approach to combustion testing. Also, they

are very useful for backing-up the more complex instruments which may develop errors and faults from time to time.

Figure 6.6, Fyrite CO₂ or O₂ indicator.

To operate the Fyrite O₂ and CO₂ indicators, a flue gas sample is extracted from an appropriate point, using the hand operated sampling assembly. The rubber cap is placed over a spring loaded plunger valve which opens when pressure is applied (**Figure 6.7**).

The aspirator rubber bulb is squeezed about 18 times in succession to clear the sampling apparatus, insuring an undiluted flue gas sample. When the plunger is released, the flue gas sample is trapped in the instrument.

The FYRITE is now inverted twice, thoroughly mixing the flue gas sample with the chemical reagents which absorb of either the O₂ or CO₂, depending on which is in use. The volume change, percent CO₂ or O₂, can be read in percent on a convenient scale.

Finding stack losses

By measuring the net stack temperature (Stack temperature - combustion air temperature), combustion tables or slide rule (**Figure 6.8**) can be used to determine flue gas losses

Measuring Flue Gas Oxygen on a continuous basis.

There are three common methods of measuring oxygen in flue gas on a continuous basis: the paramagnetic sensor, the wet electrochemical cell and the zirconium oxide ceramic cell.

Paramagnetic Sensor

The paramagnetic sensor takes advantage of the fact the oxygen molecules are strongly influenced by a magnetic field. Because of this and because other flue gas constituents, notably NO, NO_2 and certain hydrocarbons exhibit appreciable paramagnetic properties this instrument is usually limited to the laboratory.

Wet electrochemical Instruments

Wet electrochemical cells, of which there are many designs, all use two electrodes in contact with an aqueous electrolyte. Oxygen molecules diffuse through a membrane to the cathode where a chemical reaction occurs.

The electrochemical cell is essentially a battery with an electrical current that is directly proportional to the flow of oxygen through the membrane.

These cells are designed to be replaced easily, however as the flue gas sample is extracted from the stack and brought to the sensor, sample conditioning is required as well as periodic maintenance on the sampling system which becomes fouled with combustion products and a high moisture level. The high maintenance required for these sensors is a definite drawback.

Zirconium Oxide Cell

In recent years, the zirconium oxide cell has become the most common oxygen sensor for continuous monitoring of flue gases. The sensor was developed in the mid-1960s in conjunction with the U.S. space program and because of its inherent ability to make oxygen measurements in hot, dirty gasses without sample conditioning, it was quickly accepted by industrial users. The sampling element itself is a closed end tube or disk made from ceramic zirconium.

The zirconium oxide cell has several significant advantages over the other oxygen sensing methods. First, since the cell operates at high temperatures, there is no need to cool or dry flue gases before it is analyzed. Most zirconium oxide analyzers make direct oxygen measurements on the stack with nothing more than a filter to keep ash away from the cell. The cell is not affected by vibration and unlike other techniques, the output actually increases with decreasing oxygen concentration. In addition the cell has a virtually unlimited shelf life.

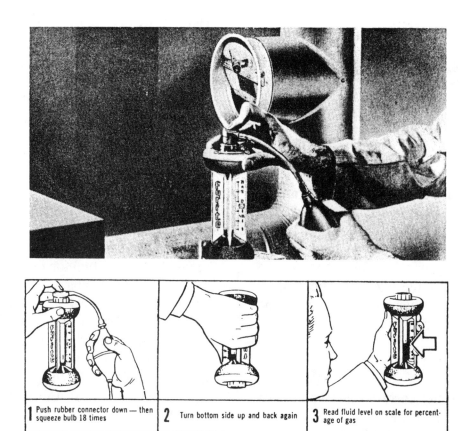

Figure 6.7, the fyrite test for CO_2 and O_2.

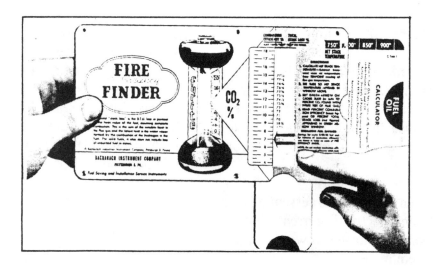

Figure 6.8, the combustion slide rule.

Insitu vs. Close-Coupled Extractive Analyzers

There are two basic types of flue gas oxygen analyzers that employ the zirconium oxide cell: insitu and closed coupled extractive.

Insitu Analyzer

In an insitu analyzer, the zirconium oxide cell is located at the end of a stainless steel probe that is inserted directly into the flue gas stream. A small heating element encompasses the cell, and a thermocouple provides feedback to an external temperature control circuit.

Close-Coupled Extractive Analyzer

A close-coupled extractive analyzer is designed somewhat differently. The zirconium oxide element and temperature controlled furnace are housed in an insulated enclosure mounted outside, but immediately adjacent to, the flue gas stack or duct.

While insitu analyzers are limited to flue gas temperatures of about 1100 F or less, close-coupled sensors can be used with high temperature probe materials up to 3,200 F.

In general, close-coupled units respond much faster to changes in the flue gas stream because they do not rely on diffusion to carry the sample to the sensing cell. A close coupled sensor can be fitted with a catalytic combustibles sensor in the same flow loop as the oxygen cell, thus making a combination oxygen/ combustibles analyzer.

Net Oxygen Vs. Gross Oxygen Measurements

As burners cannot mix fuel and air perfectly, both oxygen and unburned combustibles are in the flue gas. zirconium oxide analyzers indicate net oxygen; i.e. the oxygen left over after burning whatever combustibles are present on the hot zirconium oxide cell. Orsat, paramagnetic and wet cell oxygen analyzers measure gross oxygen.

Usually the difference between net and gross measurements are small since combustibles are generated in the parts per million range. Occasionally conditions may occur where net and gross readings are significantly different. Differences may also occur because zirconium oxide measures oxygen on a wet basis; i.e. the flue gas contains water vapor. The other measuring techniques all require cool, dry samples and are said to measure on a dry basis. The difference between wet and dry measurements can result in readings that may differ by as much as 0.5 percent oxygen.

If the condition occurs where the combustibles concentration increases to a point where there is no net oxygen in the flue gas, it becomes a sensor of net combustibles. The voltage generated by the cell increases sharply as the flue gas changes from a net oxygen to a net combustibles condition. This property of a zirconium cell is extremely useful on some combustion processes because it permits measurement on both sides of stoichiometric combustion, either excess air or excess fuel.

Relative Humidity

Relative humidity can change the amount of oxygen in air at 70 F from 20.9% at 0% RH to 20.4% at 100% RH. This 0.5% change will effect excess air settings and efficiency (0.2% or more)

Oxygen Deficiency and Safety

In measuring oxygen O_2, here are some important points to keep in mind. Normal air contains 20.9% oxygen and 79.1% nitrogen; the usual alarm point is 19.5%. Unconsciousness occurs at 15%; brain damage at 10%, and death at 5%. (These are approximate figures and may vary with the individual.

Carbon Monoxide and Combustibles Measurement Methods

There are three prevalent methods for on-line monitoring of flue gas combustibles: wet electrochemical cell, catalytic element, and non-dispersive infrared absorption.

Electrochemical Cell

The wet electrochemical cell technique is used only for carbon monoxide. It works on the principle that current flowing between the anode and cathode is directly proportional to the flow of carbon monoxide through the membrane. There are problems that occur with this type, flow rate as affected by the ambient pressure, temperature and humidity. Furthermore, the membrane can become coated with flue gas condensation, thus reducing its effectiveness. Because of this, these sensors are prone to zero and span drift.

Catalytic Combustibles Sensor

Catalytic element sensors have been widely used for detecting combustible gases in ambient air in mine shafts, parking garages, and other closed areas. High quality sensors with carefully selected elements, can be used to measure carbon monoxide in flue gas. Catalytic sensors are available with full scale ranges as sensitive as 0-2000 ppm combustibles and with accuracies of (+/-) 100 ppm or better.

The principle behind all catalytic sensors is the same, if combustibles and oxygen are both present in a gas stream, they will not normally burn together unless the temperature is elevated something above 1000 F. However, if the same gas mixture comes in contact with a solid catalyst, such as platinum, combustion will occur at temperatures as low as 400 F.

There are two elements present: One has a catalyst in an inert binder. The other is inert. The entire housing is heated to over 400 F. When the flue gas sample containing both oxygen and combustibles passes through the housing, combustion occurs on the active element but not on the reference element. This causes the temperature of the active element to rise and its resistance to change. Some close-coupled extractive oxygen analyzers have been modified to incorporate a catalytic combustibles sensor in addition to the oxygen sensor.

The catalytic sensor has the advantage of being both low cost and sensitive to hydrogen and carbon monoxide. This sensor is ideal for flue gas monitoring and recording. However, the zero and span stability of the sensor is not as good as that of the infrared sensor. Combustion systems requiring CO/combustibles measurement as an active input generally utilize infrared carbon monoxide analyzers.

Infrared CO Measurement

Carbon monoxide is one of many gases that are known to absorb infrared energy at specific discrete wavelengths. The amount of energy absorbed is a measure of the concentration of carbon monoxide.

There are two types of carbon monoxide analyzers: off-stack (sampling) and across the stack (insitu).

Off-stack CO Analyzers

Off-stack analyzers are housed in enclosures suitable for the environmental conditions and are usually located at easily accessible places near the combustion process. In most cases a sampling system is required to clean, dry, and cool the sample before it enters the analyzer. Provision for the introduction of calibration gases are usually an integral part of the design of the sample conditioning system.

Across-The-Stack CO Analyzers

Across-the-stack CO analyzers are based on the same technical principles as off-stack analyzers but their design is somewhat different. The infrared source is housed in an enclosure that mounts directly on the stack or duct. The infrared beam generated by the source passes completely through the stack into a similar enclosure mounted on the other side.

There are two major advantages of the across-the-stack systems. First, the speed of response is nearly instantaneous. Off-stack systems, conversely, can take several minutes to respond to a change in flue gas conditions. Second, across-the-stack systems provide a measurement of the average CO concentration in the stack. Unlike off-stack analyzers, which sample from a single point, they are unaffected by stratification or stagnation of flue gases in various areas of the stack.

The measurement of carbon monoxide or combustibles is an important part of achieving maximum combustion efficiency. The result will be less fuel wasted and more money saved.

Opacity

Smoking with oil and coal fuels indicates the presence of flue-gas combustibles or unacceptable flame conditions, and always should be avoided. Some boilers, especially larger ones, are equipped with smoke detectors, which can indicate poor stack conditions. Ultimately, stack conditions should be checked by visual observation.

Accurate spot check type smoke measurements can also be made with the inexpensive, portable hand pump with filter paper testers described above.

These devices use the smoke spot number or ASTM (American Society for Testing & Materials) smoke scale (standard D-2156), and can be very helpful in establishing optimum boiler conditions.

Stack Temperature

Deposits and fouling of external tube surfaces with soot, ash and other products inhibit the absorption of heat in the unit and lead to lower efficiencies. Deposits are indicated by flue-gas temperatures that are high compared to clean conditions. The efficiency loss resulting from dirty tubes can be estimated with this **RULE OF THUMB:**

EVERY INCREASE OF 40 DEG F IN STACK TEMPERATURE REDUCES EFFICIENCY BY ABOUT 1% .

Waterside deposits caused by improper water treatment also can lead to high stack temperatures, but tube failures due to overheating generally occur before any substantial efficiency losses are evident from these internal tube deposits.

Rising Stack Temperatures Indicate a Problem

Stack-temperature measurements are an easy and effective means for monitoring boiler-tube fouling. This is done by comparing the present temperature to the start up temperature or a temperature recorded when the boiler was in a clean condition.

Since stack temperatures usually increase with firing rate and excess air, make your comparisons at similar boiler operating conditions.

In the absence of previous data, flue gas temperatures normally are about 150 to 200 F above steam temperature for a boiler producing saturated steam at high firing rates.

Boilers equipped with economizers and air preheaters should be judged by observing the flue gas temperature immediately after leaving the boiler before reaching these heat recovery units.

Accuracy of Sampling Techniques.

For oxygen, CO, and smoke analyzers, the portion of the gas analyzed must be **representative** of the total gas stream. Location of the sampling site can be as important as the selection of the proper measurement device.

To illustrate: on negative draft boilers, the gas sampling point should be upstream of the air preheater, if one is installed, or upstream of any known air leaks. Reason is that air leakage into the gas ducts can dilute flue gas and resultant measurements won't give a true indication of furnace conditions. Air leakage in preheaters poses the same problem.

Sample conditions immediately downstream of bends, dampers, or induced fans should be avoided. Gases in such areas can stratify or form pockets leading to errors, especially when samples are withdrawn from a single point in a duct.

When a single-point probe is to be used, compare several readings in the duct first, to find the most representative probe location. When existing ports are not satisfactory, drill or cut out new ports and run traverse measurements. Remember, unless you get truly representative data, your testing program will be of little value.

Flue-gas temperatures are subject to stratification in ducts and a representative location of thermometers or other temperature sensors should be verified. Position them close to the boiler outlet, because thermal losses can occur in the flue gas duct, especially in uninsulated sections.

Using Carbon Monoxide to Measure Performance

On gas fired boilers carbon monoxide is the primary indicator of incomplete combustion and usually determines the lowest practical level of oxygen. The concentration of CO in the flue-gas should not exceed **400 ppm or .04%** (the limit established by many state and city ordinances, industry codes and insurance companies).

Once the final adjustments are made, it is wise to observe the operation of a boiler for an extended period to insure that your adjustments are final and there is no condition present that can increase the CO level above the acceptable limit.

When performing tests, occasional CO levels of up to 1000-2000 ppm may be encountered. Adequate boiler monitoring and flame observation are very important to assure stable conditions. Use caution at these levels because even a slight lowering of excess air can cause the CO level to skyrocket, which can lead to smoking, flame instability, furnace pulsation and possibly an explosion.

The situation is further complicated by the possibility of some CO monitoring instruments becoming insensitive and going off scale requiring a waiting period for them to come back to operating range. During this period you may be blind to what is actually happening to the CO level. One precaution that may be taken is to use a combustibles analyzer along with the CO instrument, the combustibles analyzer is less sensitive and will indicate the actual situation over a wider range.

Carbon monoxide measurements on oil and coal fired equipment is less often used because smoking or excessive carbon carryover usually precedes the formation of large quantities of CO. This is not always the case, however.

High CO levels have been measured on units where burner equipment had deteriorated or malfunctioned, impellers had burned off, oil tips plugged, overfire air was too low, etc. Also, CO can be caused by chilling the combustion process before the fuel is completely burned. Two ways this can happen is chilling of the flame with excessive (cold) concentrations of combustion air in part of the flame and through flame impingement on the (cooler) boiler tubes.

Knowing the CO level is very valuable. The CO analyzer should be capable of measuring from less than 100 over 2,000 ppm. While Orsat analyzers have traditionally been used to determine CO, difficulties in accurate reading of concentrations less than 1000 ppm have presented problems in the modern environment. Portable or permanently installed electronic type CO analyzers have the ability to measure CO continuously, having the advantage of indicating excursions in CO that may not be detected with occasional spot readings.

Chapter 7

Flow Measurement

To understand what is going on in a plant, data is necessary from the various energy streams. Flowmeters are vital for the efficient management of plants. Selecting and correctly installing the appropriate meter may take some study.

Among the things one must know about flow meters are their limitations, accuracy and rangability. The following information should make meter selection easier and insure instruments are suited to their applications.

Guide to Flow Measurement

Number	Type of Meter	Clean Liquids	Vapor or Gas	High Temperature Service
1	Orifice	1	1	1
2	Venturi	1	1	3
3	Flow Nozzles & Tubes	1	1	3
4	Pitot Tubes	1	1	3
5	Elbow	1	1	3
6	Target	1	1	3
7	Variable Area	1	1	2
8	Ultrasonic (Transit Time)	1	3	X
9	Ultrasonic (Doppler)	X	X	X
10	Positive Displacement	1	1	3
11	Magnetic	1	X	3
12	Coriolis (Mass)	1	3	3
13	Thermal (Mass)	3	1	3
14	Vortex Shedding	1	1	3
15	Fluidic	1	X	3
16	Vortex Precession	1	1	3
17	Weirs and Flumes	1	X	X
18	Turbine	1	1	3

Legend:
- (1) Designed for this service
- (2) Normally applicable for this service
- (3) Applicable under certain conditions, consult manufacturer
- (X) Not applicable for this service

Flow Measurement

1. Orifice Flow Meter (Differential Pressure Type)

Service:	Liquids, gases and steam
Design Pressure:	Determined by transmitter
Design Temperature:	Determined by materials
Flow Range:	From 1cc/minute up
Scale:	Square root
Signal:	Analog, electronic or pneumatic
Accuracy:	+/- 0.6% of max flow including transmitter; sizes smaller than 2" usually calibrated
Rangeability:	4:1 for a given transmitter span setting
End Connections:	Mounts between flanges
Sizes:	Determined by pipe size
Advantages:	Easy-to-install; uses one transmitter regardless of pipe size; low cost; wide variety of types and materials available;easy to change capacity. Versions available that do not require power.
Limitations:	Use eccentric orifices or segmental plates for very dirty fluids or slurries; quadrant orifice for viscous fluids; venturi, flow tube,pitot or elbow taps to reduce energy consumption; straight run of upstream and downstream piping required.

2. Venturi Flow Meter (Differential Pressure Type)

Service:	Liquids, gases and steam
Design Pressure:	Determined by transmitter
Design Temperature:	Determined by materials
Flow Range:	From 5 GPM liquid; 20 scfm gas; determined by pipe size
Scale:	Square root
Signal:	Analog, electronic or pneumatic
Accuracy:	+/- 1.0% of max flow or better including transmitter
Rangeability:	4:1 for a given transmitter span setting
End Connections:	Flanged
Sizes:	To 72" and larger
Advantages:	Low permanent loss; good for slurries and dirty fluids; uses one transmitter regardless of pipe size.
Limitations:	Most expensive delta-P producer; generally limited to air and water; big and heavy especially in larger pipe sizes.

Flow Measurement

3. Flow Nozzles and Tubes (Differential Pressure Type)

Service:	Liquids, gases and steam
Design Pressure:	Determined by transmitter
Design Temperature:	Determined by materials
Flow Range:	From 5 gpm liquid; from 20 scfm gas equivalent
Scale:	Square root
Signal:	Analog, electronic or pneumatic
Accuracy:	+/- 1% of full scale including transmitter; flow calibration recommended
Rangeability:	4:1 for a given transmitter span setting
End Connections:	Flange mounted or mounts between flanges
Sizes:	3" to 48"
Advantages:	Economical, low permanent loss; uses one transmitter regardless of pipe size; nozzle commonly used for steam and has higher capacity for same generated Delta-P.
Limitations:	Flow tubes lack extensive background data compared to orifice plates; application on viscous liquids limited. Calibration recommended for optimum performance.

4. Pitot (Differential Pressure Type)

Service:	Liquids and gases
Design Pressure:	Determined by transmitter
Design Temperature:	Determined by materials
Flow Range:	Determined by pipe size
Scale:	Square root
Signal:	Analog, electronic or pneumatic
Accuracy:	+/- 5.0% full scale or better including transmitter.
Rangeability:	4:1 for a given transmitter span setting
End Connections:	Insert probe
Sizes:	Unlimited probe length
Advantages:	Very low cost; uses one transmitter regardless of pipe size. Averaging types available. regardless of pipe size.
Limitations:	Doesn't sample full stream; limited accuracy. Low differential pressure for a given flow rate.

Flow Measurement

5. Elbow (Differential Pressure Type)

Service:	Liquids and gases
Design Pressure:	Determined by transmitter
Design Temperature:	Determined by materials
Flow Range:	Determined by pipe size
Scale:	Square root
Signal:	Analog, electronic or pneumatic
Accuracy:	+/- 5.0% to +/- 10% full scale including transmitter.
Rangeability:	4:1 for a given transmitter span setting
End Connections:	Mounts in 90^0 pipe elbow
Sizes:	Determined by pipe size
Advantages:	Very economical; easy-to-install; uses one transmitter regardless of pipe size;can be bi-directional; by using 45^0 tap location; very low pressure loss. Minimum upstream piping required.
Limitations:	Not good for low velocity; not as accurate as other Delta-P types. Low differential pressure for given flow rate.

6. Target

Service:	Liquids, gases and steam
Design Pressure:	Up to 10,000 psig
Design Temperature:	Up to 750^0 F
Flow Range:	0.07 gpm and up liquid; 0.3 scfm and up gas
Scale:	Square root
Signal:	Analog, electronic or pneumatic
Accuracy:	+/- 1/2% to +/- 5% full scale; factory calibrated.
Rangeability:	10:1 for any span setting; 3:1 range for any given span setting
End Connections:	Flanged, threaded, flare tube
Sizes:	Up to 8" (sampling types available)
Advantages:	No moving parts; relatively inexpensive; good for hot tarry and sediment bearing fluids.
Limitations:	Need 20 diameters upstream and 10 diameters downstream of straight pipe to maintain accuracy; reading is per cent of scale; limited range.

Flow Measurement

7. Variable Area (Rotameter)

Service:	Liquids, gases and steam
Design Pressure:	Up to 350 psig (glass tube); to 720 psig (metal tube)
Design Temperature:	Up to 400° F (glass tube); to 1000° F (metal tube)
Flow Range:	Liquids 0.01 cc/min to 300 gpm; gases 0.3 cc/min to 1500 scfm at 10 psi
Scale:	Linear or Logarithmic
Signal:	Visual, electronic or pneumatic analog
Accuracy:	+/- 0.5% of rate to +/- 10% full scale depending on type, size, and calibration
Rangeability:	5:1 to 12:1
End Connections:	Female pipe threaded or flange
Sizes:	Sizes up to 3" also used as by-pass meter around a mainline orifice for larger pipe sizes
Advantages:	Inexpensive; somewhat self cleaning; insensitive to viscosity variations below a given threshold; direct indicating; no power required; can be direct mass device; minimum piping requirements. Versions available with plastic liners.
Limitations:	Requires accessories for transmission; must be vertically mounted; gas use requires minimum back pressure.

8. Ultrasonic Transit Time (Pulsed type)

Service:	Relatively clean liquids (some designs for gas)
Design Pressure:	Wetted transducers; 1000 psig up, clamp on; pipe rating
Design Temperature:	-300° F to + 500° F
Flow Range:	Typically to 40 ft/sec
Scale:	Linear
Signal:	Analog, electronic or digital
Accuracy:	+/- 1% of rate to +/- 5% full scale depending on type and calibration
Rangeability:	Up to 40:1
End Connections:	Flanged (clamp-on design available)
Sizes:	3/8" up
Advantages:	No flow obstruction; can be bidirectional; use with any relatively clean liquid. Versions for gas. Clamp-on versions available.
Limitations:	Straight upstream piping required to provide uniform flow profile; clean liquids only.

Flow Measurement

9. Ultrasonic Doppler (Frequency shift)

Service:	Liquids with entrained gas or suspended solids
Design Pressure:	Wetted transducers; 1,000 psig up, clamp on; pipe rating
Design Temperature:	-300^0 F to $+ 500^0$ F
Flow Range:	Typically to 40 ft/sec
Scale:	Linear
Signal:	Analog, electronic or digital
Accuracy:	+/- 5% full scale or better
Rangeability:	Typically 10:1
End Connections:	Clamp-on; body versions available (sampling type available)
Sizes:	1/4" up
Advantages:	Can handle inorganic slurries and aerated liquids; clamp-on version can be installed without process shut-down.
Limitations:	Not suitable for clean liquids; requires straight upstream piping.

10. Positive Displacement

Service:	Clean liquids and gases
Design Pressure:	Pressures up to 1,400 psig for liquid and gas
Design Temperature:	up to 600^0F for liquids; up to 250^0F gas
Flow Range:	0.1 to 9,000 gpm liquid; 0 to 100,000 scfh gas. (Ultra low flow rates available)
Scale:	Linear
Signal:	Pulse or analog electronic
Accuracy:	+/- 1/2% of rate on liquid; +/- 1% of full scale on gas; factory calibrated
Rangeability:	Typically 10:1
End Connections:	Flanged or threaded
Sizes:	up to 12"
Advantages:	Ideal for viscous liquids; good for custody transfer, batching, blending; simplest versions don't require electrical power; very little straight upstream pipe required.
Limitations:	Subject to mechanical wear; requires periodic proving;sensitive to dirt and may require upstream filters; larger sizes are excessive in size and weight, may require special installation care.

Flow Measurement

11. Magnetic

Service:	Electrically conductive liquids and slurries
Design Pressure:	Up to 740 psig
Design Temperature:	Up to 360° F
Flow Range:	0.01 through 500,000 gpm
Scale:	linear
Signal:	Analog electronic or digital
Accuracy:	+/- .5% of rate to +/- 1.0% full scale; factory calibrated
Rangeability:	10:1 for any span
End Connections:	Flanged, sanitary, or wafer; Dresser and Vitaulic ends available in larger sizes.
Sizes:	0.1" to 96" (sampling type available); also used as by-pass meter around mainline orifice.
Advantages:	Unaffected by changes in fluid density, viscosity; zero head loss; bi-directional; no flow obstruction; easy to re-span; versions for dc power.
Limitations:	Moderate to expensive cost; liquids and slurries only; required minimum electrical conductivity varies with manufacturer; large sizes big and heavy.

12. Coriolis Effect (Mass)

Service:	Liquids and slurries (limited gas service)
Design Pressure:	Up to 2,800 psig
Design Temperature:	Up to 400°F
Flow Range:	Up to 23,000 LB/Min
Scale:	Linear
Accuracy:	+/- 0.5% of rate or better
Rangeability:	10:1 or better
End Connections:	Flanged or threaded
Sizes:	1/16" to 6"
Advantages:	Measures mass flow directly. Can handle very difficult applications.
Limitations:	Moderate to expensive cost. There are specific installation requirements. Head loss may be high. Be careful with two phase flows.

Flow Measurement

13. Thermal (Mass)

Service:	Gas (some designs for liquid)
Design Pressure:	500 psig and higher
Design Temperature:	Up to 150^{0}F and higher
Flow Range:	Up to 4,000 gpm liquid; up to 1500 scfm of gas
Scale:	Exponential
Accuracy:	+/- 1% of full scale
Rangeability:	10:1 or better
End Connections:	Threaded, flanged, hose
Sizes:	1/8" to 10" Sampling types available; by-pass type available.
Advantages:	Measures mass flow directly. Very low pressure loss. Good for low velocity gas measurement.
Limitations:	Affected by coatings. Some designs are fragile.

14. Vortex Shedding (Buff-body oscillatory)

Service:	Liquids, gases and steam
Design Pressure:	Pressures up to 3,600 psig
Design Temperature:	Up to 750^{0}F
Flow Range:	3 to 5,000 gpm liquid; 10,000,000 scfh gases
Scale:	Linear at high Reynolds Number.
Signal:	Frequency and analog electronic
Accuracy:	+/- 1% of rate or better on liquid; factory calibrated; +/- 2% of rate on gas.
Rangeability:	8:1 to 15:1
End Connections:	Flanged, threaded, wafer or insert; also can be used as by-pass meter around main line orifice.
Sizes:	1/2" through 8"; large sizes available. (Sampling and by-pass types available)
Advantages:	No moving parts; suitable for a wide variety of fluids; excellent combination of price and performance.
Limitations:	Straight piping required; sensitive to increasing viscosity below a given Reynolds Number.

15. Fluidic (Coanada Effect) oscillatory

Service:	Liquids
Design Pressure:	Up to 600 psig
Design Temperature:	To 250°F
Flow Range:	1 to 1000 gpm
Scale:	linear at high Reynolds no.
Signal:	Analog electronic or pneumatic; pulse
Accuracy:	+/- 1% of rate or better; factory calibrated
Rangeability:	Up to 30:1
End Connections:	Mounts between flanges
Sizes:	1" through 4"0; by-pass types available
Advantages:	No moving parts; suitable for wide variety of liquids; excellent combination of price and performance.
Limitations:	Straight piping required; sensitive to increasing viscosity below a given Reynolds number.

16. Vortex Precession oscillatory

Service:	Liquids and gases
Design Pressure:	Up to 1400 psig
Design Temperature:	-100°F to 350°F
Flow Range:	18 to 3082 GPM liquid; 10,000,000 scfh gases
Scale:	linear at high Reynolds numbers
Signal:	Frequency or analog electronic
Accuracy:	+/- 1% of rate or better; factory calibrated
Rangeability:	8:1 to 25:1 (determined by size of application)
End Connections:	Flanged
Sizes:	1/2" through 12"
Advantages:	Expensive; has operating minimum dependent on flow rate and gas density.
Limitations:	Straight piping required; sensitive to increasing viscosity below a given Reynolds number.

Flow Measurement

17. Weirs and Flumes

Service:	Liquids in open channels
Flow Range:	From 1/2 gpm
Scale:	Proportional to the measured head to the 3/2 power for rectangular and trapezoidal weirs and Parshall flumes; proportional to the measured head to the 5/2 power for "V" notch weirs.
Signal:	Analog electronic or pneumatic.
Accuracy:	2% to 5% full scale
Rangeability:	75:1 rectangular, trapezoidal weirs, Parshall flumes; 500:1 "V" notch weirs; Palmer-Bowlus flumes 10:1
Sizes:	From 1" up
Advantages:	Ideal for water and waste flows; flumes have low head loss, low cost.
Limitations:	Weirs are more accurate then flumes but require cleaning; flumes are self cleaning.

18. Turbine

Service:	Clean liquids, gases and steam
Design Pressure:	Up to 3,000 psig
Design Temperature:	Up to -450° to 500° F
Flow Range:	0.001 through 40,000 gpm liquids; to 10,000,000 scfh gases
Scale:	Linear when Reynolds number is 10,000 or higher
Signal:	Frequency or analog electronic
Accuracy:	+/- .25% of rate liquids; +/- 1% of rate gas; factory calibration should simulate operating viscosity and lubricity for liquids.
Rangeability:	10:1 to 50:1
End Connections:	Flanged or threaded
Sizes:	Up to 24" (sampling types available); also used as by-pass meter around mainline orifice.
Advantages:	One of the most accurate liquid meters; good operating range; easy-to-install and maintain; very low flow rate designs available; small in size; lightweight. Versions optimized for gas; sampling types for steam. Some versions do not require external power.
Limitations:	Sensitive to increasing viscosity; avoid use where state may change from liquid to gas; gas versions require care when used in varying flow rate applications; straight upstream pipe is required; flow straighteners may be recommended.

Chapter 8

The Control of Boilers

Controlling boilers can be a complex subject. The intent here is to summarize, simplify and clarify the subject of boiler controls. There are many very good companies specializing in boiler controls who have been continually improving control system technology for many years.

The field of boiler controls is advancing rapidly especially since the introduction of distributed digital systems, the latest in a long string of advances. However, the basics of boiler control are straightforward and subject to less change.

Steam Pressure Control

The steam pressure is the balance point between demand for energy in the distribution system and the supply of fuel and air to the boiler for combustion.

As energy is used by the steam system, pressure drops, creating a demand for more energy.

Several interacting control loops are used to insure steam pressure is maintained and that the flows of water, air and fuel are managed to insure safe, efficient and reliable operation. These control loops are:

- Boiler Safety Controls
- Combustion Control
- Feedwater Control
- Blowdown Control
- Furnace Pressure Control
- Steam Temperature Control
- Cold End Temperature Control for air heaters and economizers

Safety Controls

The purpose of the boiler safety control system is to prevent explosions and other damage to the boiler. If an accumulation of unburned fuel vapor suddenly ignites, explosive levels of energy can be released and cause severe damage to the boiler and surrounding countryside.

If the water level in the boiler drops below a certain point, it could cause a meltdown. If there are disruptive fuel pressure changes, the ensuing flame instability could cause an explosion or heavy smoking. These and several other conditions must, for the sake of safety, be controlled at all times.

Combustion Control

The energy is supplied to the boiler by the combustion process and the combustion control system regulates the firing rate by controlling amount of air and fuel delivered to the burners.

Combustion Control systems are regulated to maintain the desired steam pressure and they must be able to respond the many dynamic aspects of the burner, fuel and air control sub-systems in a coordinated way to maintain the steam pressure set point in spite of demands of the steam distribution system.

Control of Boilers

Feedwater Control

The purpose of the feedwater control system is to maintain the correct water level in the boiler during all load conditions.

Feed water control must be able to regulate water level under very dynamic conditions when the heat rate changes in the boiler causes the water level to shrink and swell, due to steam bubble volume changes in response to firing rate changes.

The boiler feed water control system must also respond to momentary changes in steam demand replacing steam that has left the boiler with feedwater.

If the feedwater control system fails, there may be serious problems. A high water level can cause severe damage to the distribution system and machinery like turbines. A low water level can allow the high flame temperatures to weaken and melt the boiler steel causing a catastrophic high energy release of steam from inside the boiler.

Blowdown Control

As water in the boiler is evaporated to produce steam, impurities are concentrated. Unless the concentration of these impurities is kept under control, severe scaling of the heat transfer surfaces and tube failure can occur.

Impurities can also be carried over into the distribution system and depending on conditions at the steam water interface, can cause surges of water into the steam system.

Blowdown is also controlled to eliminate the waste of energy by preventing more hot water than is necessary from being dumped from the operating boiler.

Furnace Pressure Control

Some larger boilers have been designed for balanced or negative draft and require a furnace pressure slightly below atmospheric for proper combustion and safe operation.

Steam Temperature Control

In boilers where superheaters are used to raise steam temperature above the saturation point, the temperature of this superheated steam must be regulated.

Cold End Temperature Control

When economizers and air heaters are used to remove heat from flue gases, sulfur oxides can form sulfuric and sulfurous acids which can cause damage to boiler and exhaust system components.

Soot Blower Control

The accumulation of soot, fly ash and other deposits on heat exchange surfaces lowers the heat transfer and efficiency of a boiler raising the stack temperature (**Figure 8.1**). It is an economic consideration to keep boilers as clean as possible.

On many boilers soot blowing operations are automated and since the accumulation of soot is roughly proportional to the number of hours the boiler has been operated, automatic sootblowers are usually activated on a simple time clock mechanism.

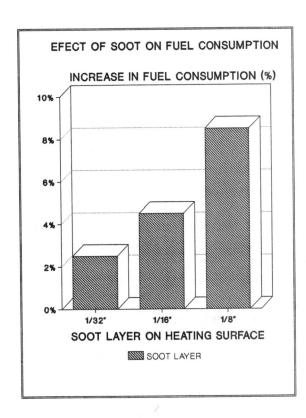

EFECT OF SOOT ON FUEL CONSUMPTION

INCREASE IN FUEL CONSUMPTION (%)

SOOT LAYER ON HEATING SURFACE

SOOT LAYER

Figure 8.1. Effect of soot on fuel consumption

Safety Controls

Safety Valves

The purpose of safety controls is to prevent explosions. Safety devices are installed for the pressure vessel which usually consists of safety valves sized to carry away steam faster than it can be generated by the boiler in the event an over pressure situation exists. This is a mechanical system consisting of safety valves and exhaust piping to vent the escaping steam harmlessly through the roof.

Boiler controls also sense overpressure and override the main control system and shut off the flow of fuel to the boiler.

Flame Safety Controls

Flame safety controls continuously monitor the pilot and main flame zone to detect the uninterrupted presence of the flame (**Figure 8.2**). If unburned fuel is allowed to accumulate in the boiler there could be an explosion, so this system must have the capability to shut down the fuel system almost immediately. These systems were developed from studying lessons learned from boiler explosions and disasters over the years.

Flame safety controls are designed to shut off the fuel to prevent explosions that can occur when:

1. An ignition source could be introduced into a furnace that contains air and accumulated fuel vapors that could become an explosive mixture.

2. Fuel is discharged into the furnace during start up without proper ignition taking place.

3. The burner flame is extinguished during normal operation without the fuel supply being shut off.

4. A major malfunction in the burner or feedwater systems.

5. Variations exist in fuel pressure which could cause flame instability.

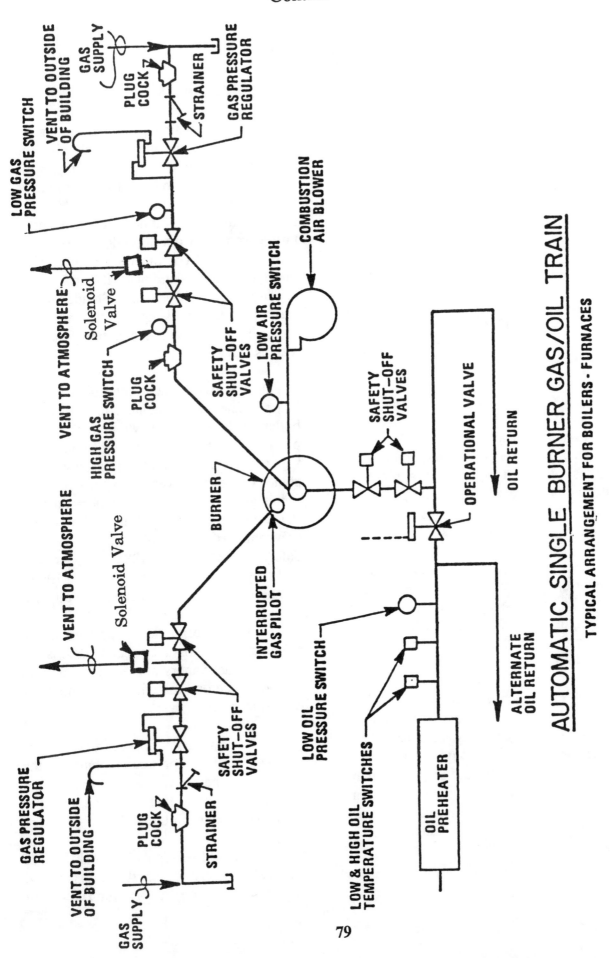

AUTOMATIC SINGLE BURNER GAS/OIL TRAIN

TYPICAL ARRANGEMENT FOR BOILERS - FURNACES

Figure 8.2, Flame safety control diagram showing safety shut off valves.

79

Control of Boilers

Purging Combustible Gases on Light-off

The start-up condition presents perhaps the greatest danger from explosion. It is essential that the furnace be purged of all combustible gases before any source of ignition is introduced and that once fuel flow is initiated, ignition take place quickly.

Purging the furnace is usually accomplished with the combustion air fan. A purging air timer does not allow ignition until the fan has been operated for a specific period of time. For watertube boilers the purge airflow must be sufficient for at least eight air changes to occur.

For firetube boilers, the purge is conducted with wide open dampers and four air changes are normally required. In either case, air flow is verified by providing limit switches on the dampers and a pressure switch at the fan discharge, or by providing airflow measurement devices such as differential pressure switches.

Once fuel has been initiated to the pilot flame or the main burner, flame verification is usually achieved by means of a flame scanning device. These flame scanners may be either the infrared flicker type or the ultraviolet type.

Recently manufacturers of flame safety control components have introduced integrated flame safety systems based on microcomputer technology with self diagnostic features. They are programmed to perform boiler start up and shutdown sequences, alarm communication and perform energy conservation functions.

Safety Interlocks

- Automatic start up sequencing has been designed to insure fuel air and combustion occur under controlled and safe conditions. Each boiler must be equipped with an approved safety system which has the capability to shut it down in the event an unsafe condition develops during start up. This system coordinates with safety interlocks for shutdown if an unsafe condition is detected. Shutdown is accomplished by stopping the start up sequence and closing the Safety Shut Off Valves (SSV).

- Loss of flame interlock; if there is no flame in the combustion zone, the boiler must be shut down immediately to prevent the accumulation of combustible vapors, which when conditions are right, could cause a powerful explosion. Pilot flame ignition, main flame ignition and main flame monitoring are all important to the prevention of combustibles build up. The Safety Shut Off Valves (SSV) close to stop fuel supply to burners when the flame scanners do not sense a flame at the burner.

- Purge interlock; the combustion products and possible accumulations of explosive fuel vapors must be purged from the boiler before each light off and after each shut down.

- Low air interlock; if air supply is less than required for safe combustion, the boiler must be shut down to prevent an accumulation of a fuel rich and possibly explosive mixture in the boiler.

- Low fuel supply interlock; loss of fuel delivery can cause flame instability, erratic combustion, poor atomization of some fuel oil equipment and the possibility of an explosion. If low fuel pressure is detected the boiler is shut down.

- Low water interlock; the combustion flame may be more than 3,000 F which is above the melting point for boiler steel. The pressure vessel is normally kept cool by the water and steam which carries away the energy from combustion. If the water level becomes low, metal temperatures can rise beyond a safe point, melting boiler components allowing a catastrophic release of energy from the steam and hot water contained in the pressure vessel.

- Other interlocks are used for specific boiler applications which override the control operation when necessary.

Basic Firing Rate Control

There are several methods of controlling steam flow, the most common method uses steam pressure to generate a master control signal. The master control signal is usually utilized by one of two types of control methods: parallel control positioning or series control positioning.

Parallel Positioning Control

Parallel positioning control is probably the most common control scheme for small industrial boilers. With this type of control the signal from the steam pressure sensor goes simultaneously to the fuel flow and air flow regulating devices. The position of these regulating devices is determined by the magnitude of the signal from the steam pressure sensor (**Figure 8.3**).

Parallel positioning systems offer the advantages of fast response and simplicity of operation and have been found to be very reliable. Individual components can be adjusted independently, so control system tuning is facilitated.

Parallel positioning systems do have a shortcoming. The master signal operates on feedback from of the actual steam pressure. The individual air and fuel regulators do not have a feedback loop to assure that the air to fuel ratios are in the desired range.

Parallel positioning control systems employ mechanical, pneumatic, electronic and digital control elements.

Jackshaft Control Systems

The jackshaft control system (**Figure 8.4**) is a simple form of parallel positioning control. **Figure 8.5** is a functional schematic of the logic of this type of system. This logic path becomes more sophisticated with the increase of boiler size where response time of different elements such as fuel and air require special consideration.

Series Positioning Control

In a series positioning control system (**Figure 8.6**), the control signal from the steam pressure sensor is not simultaneously sent to the fuel and air regulating devices. Rather, the signal is sent to only one of the regulating devices. The displacement of this first device is then measured and used to control the position of the second device. By using the actual displacement of one regulating device to control the other, series positioning control provides a margin of safety that is not available in parallel positioning control.

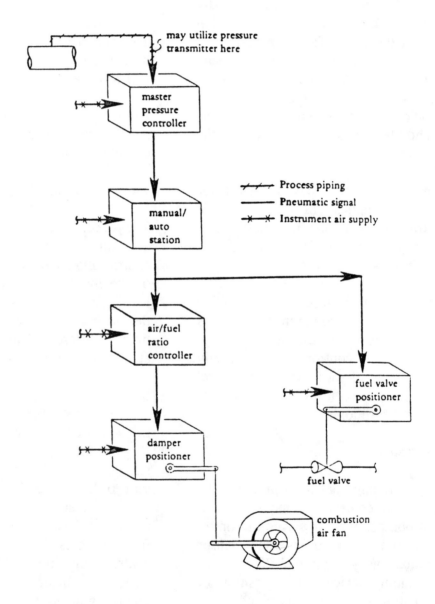

Figure 8.3. Pneumatic parallel positioning control system.

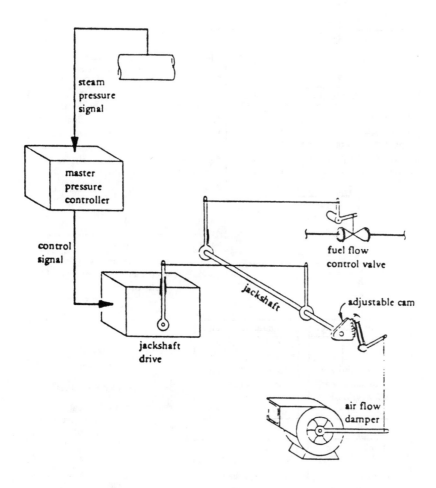

Figure 8.4. Jackshaft burner control system.

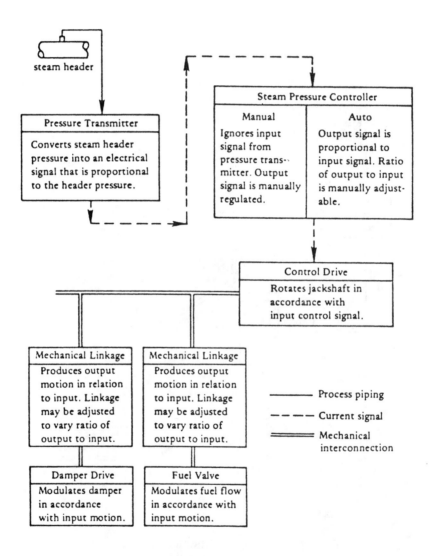

Figure 8.5. Functional schematic diagram of jackshaft system.

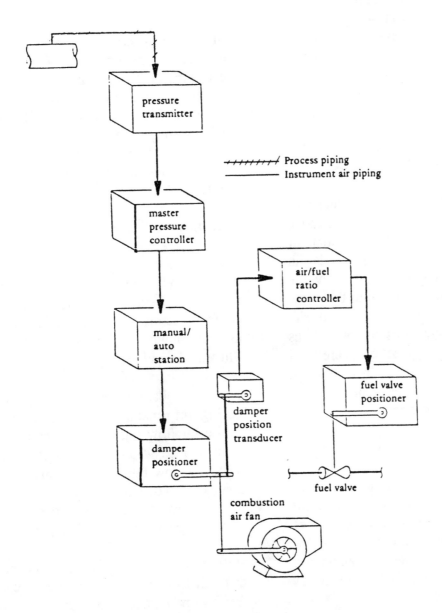

Figure 8.6. Series positioning control system (air leading fuel).

For example, if the damper positioner is the first controlled device, then the fuel flow cannot increase unless the air damper actually opens. This prevents a hazardous condition from occurring should the damper drive fail to operate correctly.

Control of the air and fuel regulators in a series positioning system is open-loop, just as in the parallel type system. A series positioning system can develop a temporary upset in the fuel to air ratio during rapid load changes. This upset can be corrected by adding "cross-limiting" or "lead-lag" to the control.

Cross Limiting Control Features

Cross limiting control prevents a fuel rich mixture on load changes. This is needed because fuel response is much faster than air system response to load changes. The lag in response of air delivery is caused by the compressible nature of air and the slow response of damper positioners and the speed change characteristics of some types of blowers. Low and high selectors are used to insure there is more than enough combustion air available to insure there is no smoking on load changes.

For example, the high selector receives the fuel demand signal and actual fuel flow signal; it sends the higher value to the air damper as the set point to for air flow. On the other hand, the low selector compares the fuel demand signal and the computed fuel signal that the air flow can handle and sends the lower one to the fuel flow valve as the control point.

Metering Control Systems

Metering control systems overcome both the shortcomings of parallel and series positioning control, but the added cost and complexity of metered control has generally restricted their use to large boilers (**Figure 8.7**).

With metering control, the control of the fuel and air regulators is a closed loop. The steam pressure is measured and feedback is provided to the master controller which adjusts the fuel and air flows. The fuel and airflows are also measured and feedback is provided to their control devices to insure they are in accord with the master controller.

Feedwater Control

The purpose of feedwater control is to maintain the correct water level in the boiler during all load conditions. The simplest control systems use one variable, water level as the input to the control system as illustrated in **Figure 8.8**.

The controllability can be improved by adding information about steam flow and feedwater flow rates. This is called a three element feed water control system (**Figure 8.9**), the flow rates of steam and water are compared to overcome any momentary false water levels caused by the shrink and swell affects due to rapid load changes.

With the "shrink-swell" phenomena it is difficult to measure drum level accurately. As steam demand increases, the steam pressure decreases lowering drum pressure. This decrease in pressure plus the increased firing rate to meet this higher demand

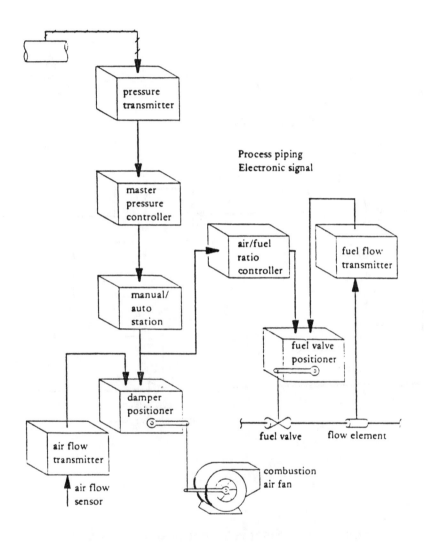

Figure 8.7. Metering control system.

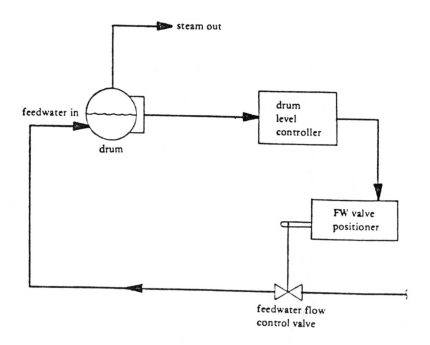

Figure 8.8. Single element feed water control.

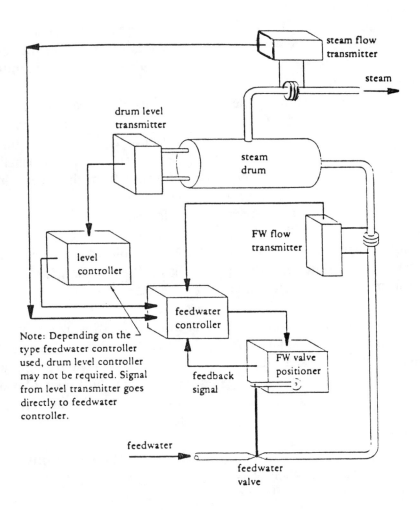

steam flow
transmitter

steam

drum level
transmitter

steam
drum

FW flow
transmitter

level
controller

feedwater
controller

FW valve
positioner

Note: Depending on the
type feedwater controller
used, drum level controller
may not be required. Signal
from level transmitter goes
directly to feedwater
controller.

feedback
signal

feedwater

feedwater
valve

Figure 8.9. Three-element feed water control.

increases the volume of steam bubbles in the drum water, increasing its water level temporarily. This swelling of water level sends a false signal to the drum level controller indicating a need for less water rather than more, as the circumstances actually require. Conversely, as steam demand decreases, drum pressure increases and there is less combustion heat for steam formation. This contracts the volume of steam bubbles in the drum water which the drum level controller interprets as a need for more water, which is colder than the boiler saturation temperature causing further shrinking.

The three element control loop is very precise and stable because of the closed-loop control of the feedwater valve. It measures steam flow and water flow as well as drum water level. This system can adjust for actual steam and water flow conditions balancing this information against measured water level.

Feedwater controllers are available in pneumatic, analog electric and digital electronic models.

Blowdown Control

Blowdown is done either intermittently or continuously. On many smaller boilers blowdown is performed manually after an analysis of the boiler water indicates that the water is exceeding impurity limits.

If too much hot boiler water is blown down from the boiler an excessive amount of energy is wasted. If the concentration of impurities in the boiler is allowed to get to high, carry-over and scale formation can occur.

Control systems are available which will automatically perform the operations necessary to blow down the boiler. Automatic blowdown controllers use electrochemical cells to measure the concentration of minerals by measuring electrical conductivity and acid (PH) levels.

The signal corresponding to the impurity level is compared to the setpoint and when exceeded, blowdown is initiated.

Boilers operated at a steady load may be equipped with a continuous blowdown system. A continuous blowdown system removes a continuous, usually small amount, of water from the steam drum to keep the concentration of undesirable chemicals at a safe level.

Manual blowdown to remove sludge which has settled in the mud (lower) drum is still necessary.

Boiler blowdown water is often passed through a heat exchanger to recover the energy which would otherwise be wasted. Heat from the boiler blowdown is usually used to heat the incoming boiler makeup feedwater.

Furnace Pressure Control

Furnace pressure conditions must be closely controlled to establish safe and efficient firing conditions. Air fuel mixing for the burner systems depends to a great degree on the differential pressure across the burner.

Different challenges apply to different sizes of boilers. **Figure 8.10** shows natural draft, balanced draft, induced draft and forced draft systems. Each configuration will require special control considerations.

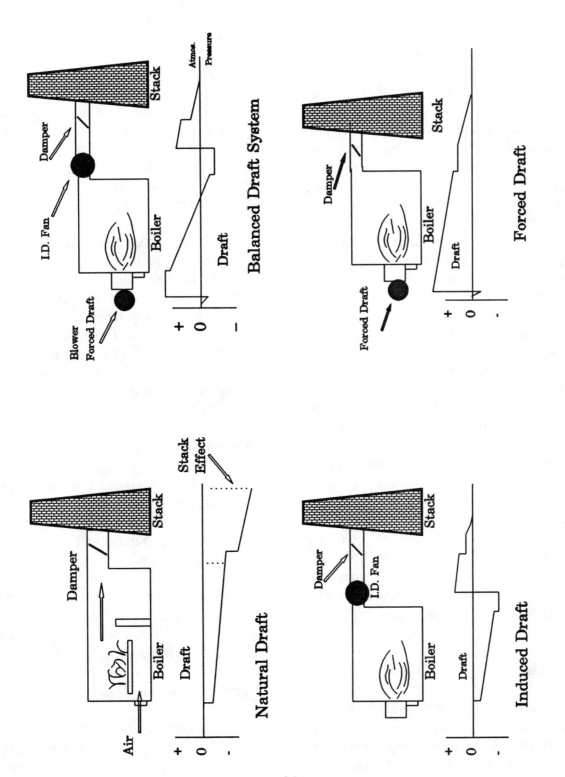

Figure 8.10. (a) Natural Draft, (b) Balanced Draft, (c) Induced Draft and (d) Forced Draft systems.

Damper openings, forced draft fans, induced draft fans and stack effect all play an important role in various furnace pressure control systems.

Steam Temperature Control

Superheaters are used to raise the steam temperature above the saturation point for superheated steam. In doing so, the temperature of the superheated steam must be regulated. The main control loop is to regulate the superheated steam temperature by a desuperheater steam valve. This control loop is enhanced by a feed forward, usually air flow, signal into the loop plus an inner loop cascade signal for spray water flow to eliminate the effects of spray valve performance variations.

Cold End Temperature Control

Cold End Corrosion occurs when metal temperatures fall below the dewpoint for sulfurous and sulfuric acids. The most critical point for the economizer is the feedwater entrance point to the heat exchanger where metal surfaces can be cooled below the acid dew point.

In air preheaters a steam or glycol heating system is used to regulate the cold end temperature. A single element control of the heating control valve may be adequate. However, the three element air heater cold-end temperature control may have to be used.

Controls to Improve Efficiency

To a great extent, the efficiency of a boiler is dependent on the design of the burner, heat exchanger and other design parameters which cannot be easily changed. However, changes in fuel composition, air temperature, pressure and humidity, boiler load and equipment condition all introduce changes in boiler performance which can be accommodated for improved control.

In general "improved control" means better control of the air fuel ratio. The simple parallel positioning and series positioning systems have open-loop control of the fuel and airflows. There is no feedback into the control system which tells the system what the actual fuel and airflows are. Feedback is provided only by the steam header pressure.

If the header pressure is at the control set point but the fuel to air ratio is nowhere near the desired value, the control system will take no action to correct it. For this reason it is highly desirable to modify a boiler control system to include feedback of information on the air to fuel ratio. This is done by adding either an oxygen trim or carbon monoxide trim system to the boiler controls.

Oxygen Trim

The idea behind oxygen trim controls is to maximize boiler efficiency by operating at the point where the combined efficiency losses due to unburned fuel and excess air losses is minimized.

An oxygen trim system measures the excess oxygen in the combustion products and adjusts the airflow accordingly for peak combustion efficiency (**Figure 8.11**).

Carbon Monoxide Trim

Carbon monoxide (CO) trim systems are also used to excess air. Carbon monoxide trim systems, in fact offer several

92

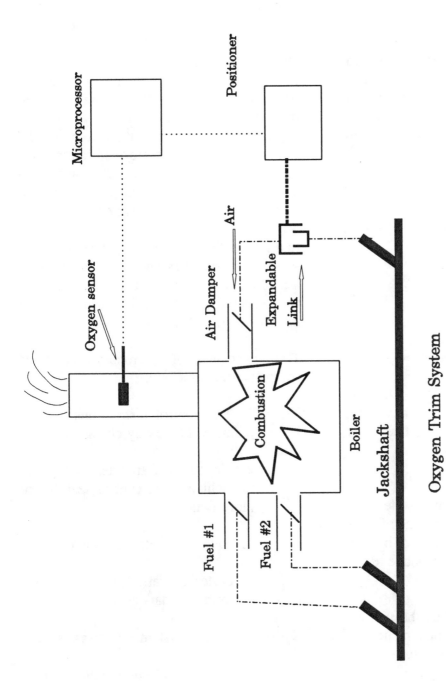

Figure 8.11. A jackshaft control system with oxygen trim control.

advantages over oxygen trim systems.

In the carbon monoxide system trim system, the amount of unburned fuel (in the form of CO) in the flue gasses is measured directly and the air to fuel ratio control is set to actual combustion conditions rather than pre-set oxygen levels. This way the carbon monoxide trim system is continuously searching out the point of maximum efficiency **(Figure 8.12)**.

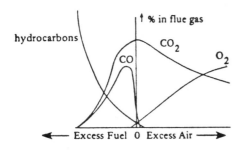

Figure 8.12. Variation of flue gas constituents with excess air level.

Other benefits of Carbon monoxide trim systems are that it is independent of fuel type and it is almost unaffected by air infiltration common in the negative draft type of boilers.

Carbon monoxide systems must be viewed with caution because the carbon monoxide level is not always a measure of excess air. A dirty burner, poor atomization, flame chilling, flame impingement on boiler tubes and poor fuel mixing can also cause a rise in carbon monoxide level.

Direct Digital Control

The latest developments in automatic boiler control systems have been in the technology of direct digital control (DDC) of boilers. DDC systems are based on microcomputer technology and control the boiler by means of "software" rather than fixed interconnections or "hardwired logic." Digital control systems often combine the functions previously performed by separate hardware systems such as combustion control, safety controls and interlocks, and monitoring and data acquisition **(Fig 8.13)**.

In a typical DDC system, input devices such as sensors, switches and position encoders are accessed at very short intervals insuring any problem is dealt with almost immediately.

The advantages of microcomputer based control systems include:

1. The control instructions and set points can be easily changed.

2. The systems are easily expanded to include more control features or more boilers.

3. Low cost control redundancy.

4. Boiler operation diagnostics and performance data acquisition.

5. Control system self-diagnostics.

6. Easy access to important real-time or historical information by telecommunication links.

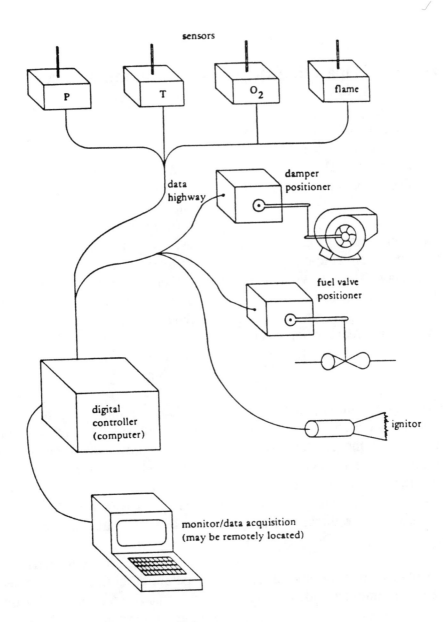

Figure 8.13. Digital control system.

Chapter 9

Boiler Tune Up

The Importance of Operating Boilers with Minimum Excess Air

Reducing excess air is one of the most effective boiler improvement techniques one can apply without high capital cost. When excess air is reduced, several things are accomplished:

- When hot combustion gases leave a boiler, they have the potential to carry away a lot of waste energy. The less volume of exhaust gas, the less loss there is.

- Flue gas velocity is reduced increasing the time available for heat transfer in the boiler.

- Flame temperature is raised, increasing radiant heat transfer in the combustion zone walls. Heat transfer in the combustion or radiant heat transfer zone is very efficient and becomes more efficient as flame temperature goes up. This increase of heat exchange efficiency reduces stack temperatures.

- Pollution is reduced because less fuel is required to meet the same demands.

The increase in efficiency available from tuning up a boiler has three direct and related benefits: (1) It Saves fuel dollars, (2) it reduces the cost of energy at the point of use and (3) it increases the steam output available from a boiler (increased productivity).

When evaluating boiler efficiency improvement projects, cost and benefit calculations must be based on the tuned up efficiency of the boiler to prevent false estimates of benefits. It doesn't make sense to attempt to correct a problem by adding something new to a boiler if it can be corrected by maintenance and repairs and a tune up.

A Tune Up Starts With an Inspection and Testing.

Efficiency improvements obtained under a deteriorated state of the boiler can be substantially less than the improvements achieved under proper working conditions. Therefore, it is essential that the boiler be examined prior to testing and that necessary repairs and maintenance be completed.

One of the first questions in tuning up an operating boiler is whether or not is necessary to take the boiler out of operation and go through the expense of opening it up for a formal inspection.

A preliminary efficiency test and review of records might provide valuable information about a boiler's operating condition and whether or not a more detailed inspection is necessary.

The condition of the burner system and combustion process can be judged by the excess oxygen level. Boiler start up records and records of previous tune ups provide a valuable reference point for your tune up program. Even a call to the manufacturer can provide useful information on the boiler's expected performance characteristics and minimum expected excess air levels (**Fig 9.1**).

If this information is not available, then general information can be used based on information from typical minimum oxygen settings in similar boilers. The following general information on minimum excess oxygen is based on a large number of boiler tests and is applicable to high firing rates. As firing rate decreases burner performance falls off and more excess air may be needed for some burners.

- For natural gas boilers, 0.5% to 3%.

- For liquid fuels, 2% to 4%.

- For pulverized coal, 3% to 6%.

- For stoker fired coal, 4% to 8%.

Stack Temperature

Stack temperature measurements are an easy and effective means for monitoring boiler tube cleanliness and the general effectiveness of the heat exchange process in a boiler. Existing temperatures can be compared to values obtained during start up or after maintenance and cleaning, to identify any deviations from baseline levels. Since stack

temperature usually increases with firing rate and excess air, make your comparisons at similar boiler operating conditions.

If previous information is not available or if temperatures seem excessive, the following graph (**Fig 9.2**) can be used for general estimates. Temperature readings to measure boiler performance must be taken before economizers or air heaters cool down the flue gases   but overall performance is judged from the temperature after these units.

Inspecting Your Boiler

Boiler components that you should inspect before conducting efficiency tests include burners, combustion controls and furnace. Typical things to look for are shown below (**Fig 9.3 through 9.5**). Consult your boiler manufacturer for a more complete list appropriate to the specific equipment in your plant.

Oil burners

- Make sure the atomizer is suitable for your present firing conditions; for the type of oil being burned and burner geometry.

- Verify proper flame pattern through the viewing ports located at the sides and back of the boiler, if installed.

- Inspect burners for warping or overheating, coke and gum deposits. Clean or replace parts as appropriate.

- Inspect oil tip passages and orifices for wear and scratches or other marks. Use proper size drill   or machine gages for testing.

Combustion Efficiency
Oxygen-CO Relationship

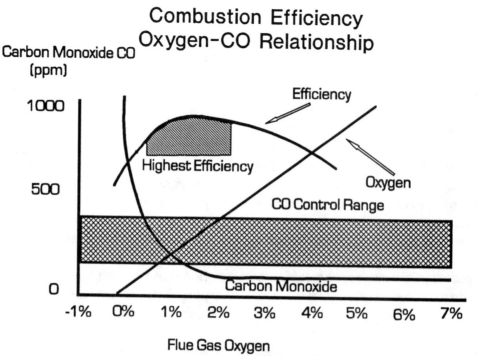

Figure 9.1, CO-Oxygen relationship showing CO control range and the effect of CO and oxygen levels on combustion efficiency.

Flue Gas Temperatures
Above Steam Temperature

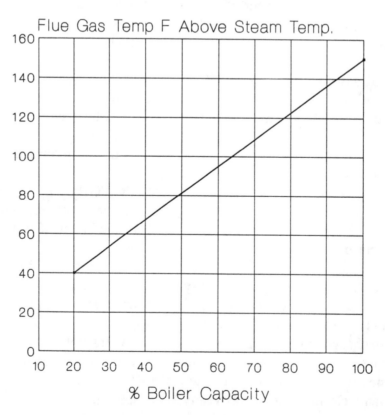

Figure 9.2, Boiler exhaust temperature above temperature of water-steam temperature in boiler.

Gas Fired Burners	Condition and cleanliness of gas injection orifices.
	Cleanliness and operation of filter & moisture traps.
	Condition of diffusers, spuds, gas cans, etc.
	Condition of burner refractory.
	Condition and operation of air dampers.
Oil Fired Burners	
	Condition and cleanliness of oil tip passages.
	Oil burning temperature.
	Atomizing steam pressure.
	Condition of impeller/diffusers.
	Position of oil guns.
	Cleanliness of oil strainer.
	Condition of burner throat refractory.
	Condition and operation of air dampers.
Pulverized Coal Firing	
	Condition and operation of pulverizes, feeders and conveyors.
	Condition of coal pipes
	Coal fineness
	Erosion and burnoff of firing equipment.
	Condition and operation of air dampers.
Stoker Firing	
	Wear on grates
	Position of all air proportioning dampers.
	Coal sizing
	Operation of cinder reinjection system

Figure 9.3 Preliminary Burner Equipment Checklist I

Combustion controls	Cleanliness and proper movement of fuel valves
	Smooth repeatable operation of all control elements.
	Adequate pressure to all regulators.
	Unnecessary cycling of firing rate
Flame Safety System	
	Proper operation of all safety interlocks and boiler trip circuits.
Furnace	
	Excessive deposits or fouling of gas-side boiler tubes.
	Proper operation of sootblowers.
	Casing and duct leaks.
	Clean and operable furnace inspection ports.

Figure 9.3 [II], Preliminary Boiler Inspection Checklist II

COMMON CAUSES OF LOW CO₂ AND SMOKY FIRE ON OIL BURNERS

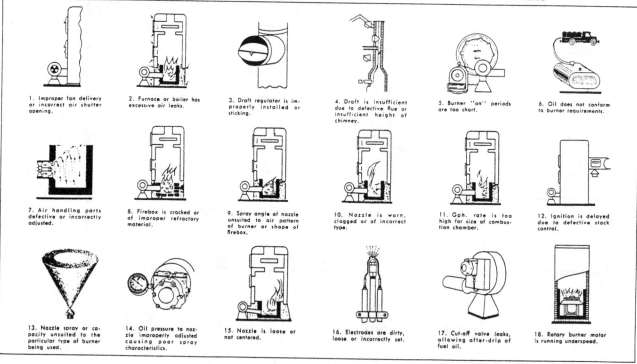

Figure 9.4 Troubleshooting guide for small oil burners. (Bacharach Co.)

● Verify proper oil pressure and temperature at the burner. This may include resetting the fuel oil heater to the grade of oil presently being fired, calibrating gages and instruments and resetting pressures to manufactures values or values indicated in engineering records. If changes are made, careful observation of actual effects they produce is important in case a problem develops.

● Verify proper atomizing-steam pressure. Also, be aware a defective steam trap may introduce unwanted water into the flame zone.

● Make sure that the burner diffuser (impeller) is not damaged, and is properly located with respect to the oil gun tip.

● Check to see that the oil gun is positioned properly within the burner throat, and the throat refractory is in good condition.

Gas burners

● Be sure that filters and moisture traps are in place, clean, and operating properly to prevent gas orifice plugging. Inspect gas-injection orifices and verify that all passages are unobstructed.

● Look for any burned off or missing burner parts. Confirm location and orientation of all parts. (Viewing ports are very helpful to identify faulty flame patterns or other problems)

Pulverized-coal burners

● Verify that fuel- and air- components, pulverizers, feeders, primary and tempering-air dampers etc. are all working properly.

● Clear coal pipes of any coal and coke deposits.

● Check burner parts for any signs of excessive erosion or burn-off.

Spreader-stoker firing

● Check grates for wear. Check the stokers and the cylinder- reinjection system for proper operation.

● Confirm the proper positioning of all air-proportioning dampers.

● Verify proper coal sizing.

Combustion controls

● Be sure that all safety interlocks and boiler trip circuits operate.

● See that all system gages are calibrated and functioning.

● Eliminate play in all control linkages and air dampers. Also check to see if there is accurate repeatability when load points are approached from different directions.

● Check control elements for smooth accurate operation. Correct unnecessary hunting caused by improperly adjusted regulators and automatic master controllers.

● Inspect all fuel valves to verify proper movement, clean and repair as necessary.

Furnace

● The firesides should be clean, check for sootblower cleaning efficiency. Consider periodic water-washing if firesides are not being kept clean by normal soot-blowing.

• Inspect and repair internal baffling. Defective baffling allows hot combustion gases to escape without giving up heat causing high stack temperatures. A traverse of the breaching with a temperature indicator may point out local hot spots behind baffle defects. Once hot spots are identified, the defects can be corrected.

• Repair any casing leaks and any cracked or missing refractory.

• Clean furnace-viewing ports and make sure that burner throat, furnace walls and leading convection passes are visible. Being able to see the condition of the flame, burner, refractory zone and furnace is essential to detecting and correcting problems.

Flame appearance

The flame is the heart of the combustion process; if it isn't right you will have a serious challenge tuning-up a boiler.

The appearance of a boiler's flame offers a good preliminary indication of combustion conditions. It is difficult to generalize the characteristics of a "good" flame because of the variations due to burner design and operating conditions.

As the ideal situation is to operate with low-excess air, one must be familiar with the conditions this will create compared to higher excess-air conditions which may be favored by operators. Low excess-air operations demands that plant personnel pay close attention to the combustion process.

• Reduced oxygen levels leads to increased flame length because it takes more time to burn completely. It actually grows in size, filling the furnace more completely.

• It exhibits a lazy rolling appearance. Instead of intense, highly turbulent flames, low-oxygen flames may appear to move somewhat more slowly through the furnace.

• It has an over-all color that may change as excess oxygen is decreased. Natural gas flames for instance, become more visible or luminous with yellow or slightly hazy, portions. Coal and oil flames become darker yellow and orange and may appear hazy in parts.

Although low excess-air operation is important, it is sometimes not possible to operate this way because of combustion related problems.

Observing oil flames provides important information concerning the combustion process. The combustion problems which typically occur will be due to one or more of the following.

1. Excess oxygen level.

2. Oil temperature or pressure.

3. Oil gun tip.

4. Air register setting.

5. Oil gun position.

Listed below are certain problems that may come about in a well adjusted flame.

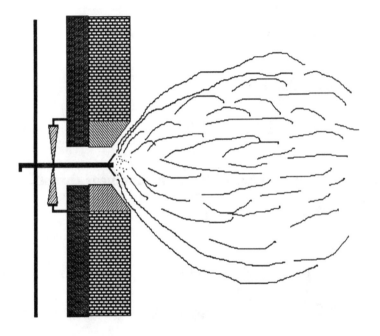

Figure 9.5 shows the flame pattern for a nearly ideal flame geometry.

The ideal flame:

- Flame barely clears burner throat.

- Flame is bright and smooth with light smoke wisps on the end.

- Problems, none.

103

Problems that can occur in a burner flame of ideal geometry.

Flame contains ragged sparks

- Excess oxygen too high
- Oil too hot.
- Ash in fuel.

Flame very smokey looking

- Excess oxygen too low.

Flame black and oily-looking with no burning near throat.

- Air register closed.

Flame appears to be less bright and dense, somewhat transparent.

- Oil pressure delta-p too high. Too much return flow.
- Worn tip returning too much oil flow.

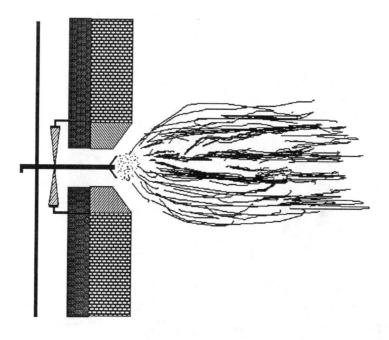

Figure 9.6. Flame is narrow and does not fill the throat. Possible thin stream of oil present in the middle of the flame.

Problem:

- Low oil pressure delta-p or return oil line plugged.

- Air register open too far.

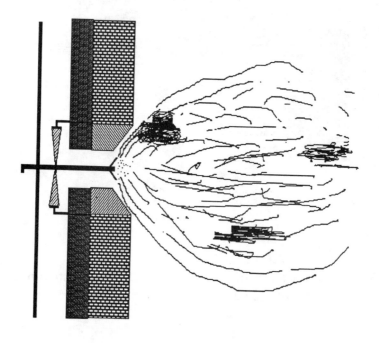

Figure 9.7, Flame pattern with intermittent oil slugs coming out of the tip.

Problem:

- Atomizer problem, tip partially plugged, tip worn or other burner tip problem.

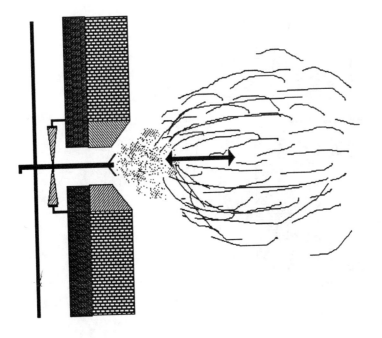

Figure 9.8, Flame blow off. May be continuous or pulsating on and off the tip.

Problem:

- Air register open too far.

- Oil gun positioned too far in toward the furnace.

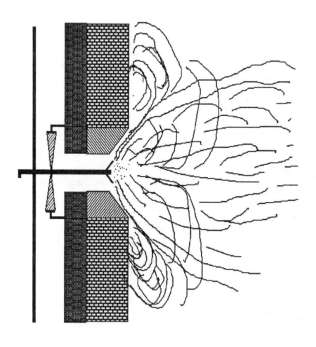

Figure 9.9. Flame clears throat, but rolls back, impinges and rolls up the furnace wall.

Problem:

- Air register closed too far.

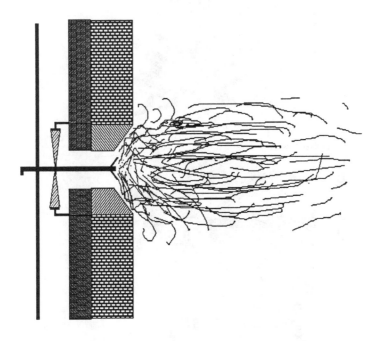

Figure 9.10. Flame impinges on burner throat.

Problem:

- Oil gun not fully extended into firing position.

- Wrong tip or worn tip orifice.

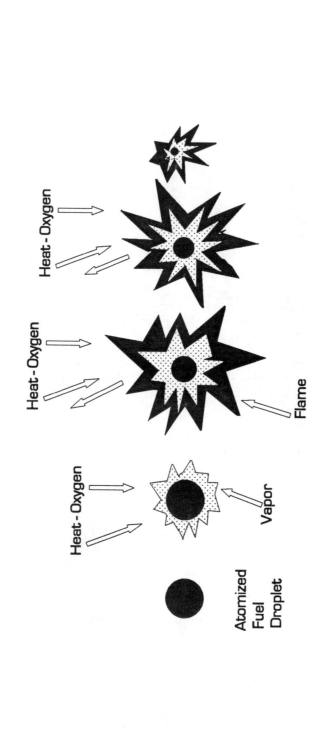

Figure 9.11. Combustion of Oil: (a) the oil is broken down into fine spherical particles by the atomization process to develop as much surface area as possible for the formation of a vapor for combustion, the vapor burns not the oil itself. Next, (b) The intense heat in the combustion zone heats the vapor to the ignition point and oxygen is mixed with the fuel to support combustion, (c) The vapor and carbon residue continues to burn reducing the size of the particle as it moves away from the burner, and (d) all fuel should be burned in the combustion zone to insure a clean stack and no build up within the boiler.

Problem	Probable Cause	Corrective Action
Heavy White Plume • Consistent	High excess oxygen	Lower excess oxygen down to a level slightly above smoke limit.
	High sulfur fuel	Have fuel sample analyzed.
• Transient	Rapid increase in load	Check cross limiting controls. Make load changes more gradual.
Black smoking • At low excess oxygen	Insufficient air	Increase air. Check operation of combustion controls when in automatic mode. Sluggish control response on load swings will cause occasional smoking. Automatic (load control) operation will usually require a slightly higher oxygen setting.
• At high excess O_2 and low wind box-furnace pressure, flames look about the same throughout the furnace; slow and lazy indicating low excess air.	O_2 instrument out of calibration	Check calibration of O_2 instruments.
• At high excess O_2 and high windbox-furnace pressure, flames look about the same throughout the furnace.	Oil temperature problems Oil pressure problems	Check oil heater and oil temperature at burner. Check oil supply-return pressure. Check pressure gage calibration.
• At high excess O_2 and high windbox-furnace pressure, flames are not the same throughout the furnace.	Localized combustion problem.	Check flame patterns 1. Oil temperature 2. Oil supply-return pressure 3. Oil gun position If these are ok, then check air register settings and operation, reposition if necessary. Problem may be burner tip related, inspect and change if necessary.
• Excess O_2 cycling with constant forced-draft-fan and fuel flow indications.	Localized air heater pluggage	Check air heater pressure drop.
• Excess O_2, Forced-draft-fan and fuel flow cycling	Control problem	Check combustion control system.

Table 9.1. Trouble Shooting Performance Problems

The Basics of a Tune Up

Keep in mind, the basic criteria for good combustion:

- Time
- Temperature
- Turbulence
- Sufficient air

Note: These tests and adjustments should only be conducted with a through understanding of the test objectives and following a systematic, organized plan.

Instruments

The minimum limits of excess air should be approached cautiously with flue gas analyzers which continuously provide an accurate measurement of average conditions for the burner being adjusted. When the maximum smoke spot number, for oil or the maximum carbon monoxide level for natural gas is reached it should be noted along with the corresponding burner settings. Flame instability is sometimes a limiting factor on reducing excess air. In stoker fired boilers, the onset of clinkering or overheating of the grate sometimes precedes the formation of smoke or carbon monoxide.

Safety

- Extremely low excess oxygen operation can result in catastrophic results.

- Know at all times the impact of the modification on fuel flow, air flow and the control system.

- Observe boiler instrumentation, stack and flame conditions while making any changes.

- When in doubt, consult the plant engineering personnel or the boiler manufacturer.

- Consult the boiler operation and maintenance manual supplied with the unit for details on the combustion control system or methods of varying burner excess air.

Finding the Smoke and CO Threshold

Once your boiler is in good working order, the next major step in improving efficiency and reducing emissions is to establish the lowest level of excess oxygen at which the unit can operate safely and meet clean air laws.

Since most boilers operate over a reasonably broad range, tests must be run at several firing rates to determine the minimum excess-oxygen level for each. Only then can the combustion-control system be tuned for optimum fuel economy.

At each firing rate investigated, excess-oxygen in the flue gas should be varied from 1-2% above the normal operating point down to where the boiler just starts to smoke, or to where CO emissions vary between 150 to 250 Parts per million (PPM). The level of 400 PPM is the legal limit established in many states and by insurance companies. This condition is referred to as the smoke or CO threshold, or simply as the minimum oxygen point.

The smoke threshold generally applies to coal and oil firing, because smoking usually occurs before CO emissions reach significant levels. The CO level pertains to gaseous fuels. The smoke threshold for solid and liquid fuels represents the lowest possible excess-oxygen level at which acceptable stack conditions can be maintained.

The Smoke Spot Number (SSN) is a scale of smoke density which can be related to the soot accumulation in a boiler. **Figure 9.12** shows the desirable SSN for various fuels. **Figure 9.13** shows the relationship of SSN to the rate at which deposits generally accumulate. The SSN is directly related to excess-oxygen levels and burner performance.

Minimum Excess Oxygen

A proven method for determining the minimum amount of excess oxygen required for combustion involves developing curves similar to the smoke/oxygen and CO/oxygen curves shown in **Figures 9.14 and 9.15**. Based on test measurements, these curves show how boiler smoke and CO levels change as excess oxygen is varied.

Each of these figures depicts two distinct curves, illustrating the extremes in smoke and CO behavior that may be encountered. One curve exhibits a very gradual increase in CO or smoke as the minimum excess oxygen condition is reached. The other has a gradual slope at relatively high oxygen levels and a steep slope near the maximum oxygen point. For cases represented by this second curve, unpredictably high levels of smoke and CO, or potentially unstable conditions, can occur with very small changes in excess air.

Caution is required when reducing air flow near the smoke point or CO threshold. Carefully monitor instruments and controls, flame appearance and stack conditions simultaneously. Decrease the level of excess oxygen in very small increments until you find out what kind of a curve is developing. It can be steep and possibly unstable or gradual. Some boilers have a gradual characteristic at one firing rate and a steep characteristic at another.

Step by step Procedure for Adjusting Boiler Controls for Low-Excess Oxygen

1. Establish the desired firing rate and switch combustion controls from automatic to manual operation. Make sure all safety interlocks are still functioning.

2. Record boiler and stack data (pressure temperature etc.), and observe flame conditions after the boiler operation stabilizes at the particular firing rate selected. If you find that the amount of excess oxygen in the flue gas is at the lower end of the range of typical minimum values and the CO and the smoke are at acceptable levels, the boiler is already operating at a near optimum air to fuel ratio. This may not be so at other firing rates. It may still be desirable, however to complete the remaining portion of this procedure, to determine whether still lower levels are practical.

3. Increase air flow to the furnace until readings of excess oxygen at the stack increase by 1-2%. Again, be sure to take readings after boiler operation stabilizes and note any changes in flame conditions.

4. Return air flow to normal level and begin to slowly reduce it further, in small increments. Watch the stack for any signs of smoke and constantly observe the flame and stack. Record stack excess-oxygen reading, smoke spot number, the concentration of CO in the flue gas and the stack temperature

Maximum Desirable Smoke Spot Number	
Fuel Grade	Maximum Desirable SSN
No. 2	Less than 1
No. 4	2
No. 5	3
No. 6	4

Figure 9.12. Maximum desirable Smoke Spot Number for various fuels.

Smoke Spot Number	EFFECT OF SMOKE ON SOOT BUILD UP	
	Rating	Sooting Produced
1	Excellent	Extremely light if at all
2	Good	Slight sooting which will not increase stack temperature appreciably.
3	Fair	May be some sooting but will rarely require cleaning more than once a year.
4	Poor	Borderline condition. Some units will require cleaning more than once a year.
5	Very Poor	Sooting occurs rapidly and heavily

Figure 9.13. The effect of Smoke Spot Number on soot build up.

Combustion Efficiency
Oxygen-Smoke limit Relationship

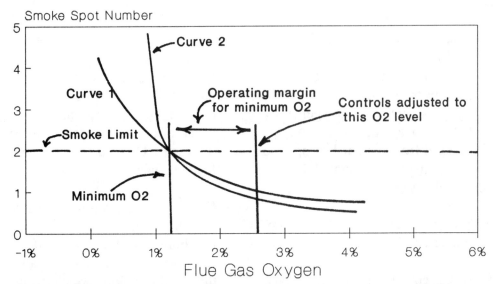

Figure 9.14. Characteristic curve identifies minimum excess air and tune-up control settings for oil fired boilers. **Curve 1** Gradual smoke/O_2 relationship. **Curve 2** Steep smoke/O_2 relationship. The type of curve is dependent on burner operating characteristics and varies with the firing rate and the particular type of burner.

Combustion Efficiency
Oxygen-CO Relationship

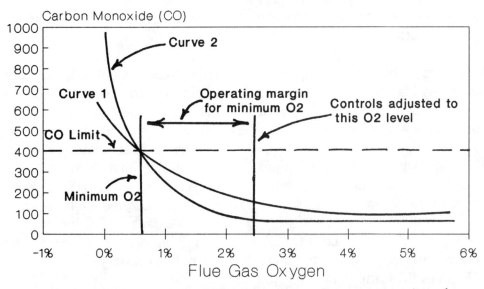

Figure 9.15. Characteristic curve identifies minimum excess air and tune-up control settings for gas fired boilers. **Curve 1** Gradual carbon monoxide/O_2 relationship. **Curve 2** Steep carbon monoxide/O_2 relationship. The type of curve is dependent on burner operating characteristics and varies with the firing rate and the particular type of burner.

after each change.

Do not reduce air flow by throttling the burner air registers, because this alters the fuel air mixing characteristics and complicates the tests. Also, if you run tests at low firing rates, which is not generally recommended, keep a close watch on the windbox/furnace differential. If it drops too low, a fuel trip may be activated by the burner safeguard system.

5. Continue to reduce the airflow step-wise until you reach one of these limits:

- Unacceptable flame conditions-such as flame impingement on furnace walls or burner parts, excessive flame carryover, or flame instability.

- High level of CO in the flue gas.

- Smoking at the stack. Do not confuse smoke with water vapor, sulfur or dust plumes which are usually white or gray in appearance and remember to observe local air pollution ordinances.

- Incomplete burning of solid fuels. Recognize this by high carbon carryover to dust collectors or increased amounts of combustibles in the ash.

- Equipment-related limitations such as low windbox/furnace pressure differential, built-in air-flow limits, etc.

6. Develop O_2/smoke or O_2/CO characteristic curves, similar to those shown in **Figures 9.14 and 9.15** using the excess oxygen and CO or smoke-spot number data obtained at each air-flow setting.

7. Find the minimum excess-oxygen level for the boiler from the curves prepared in step 6, but do not adjust the burner controls to this value. Though this may be the point of maximum efficiency, as well as minimal NOX emissions, it usually is impractical to operate the boiler controls at this setting, because of the tendency to smoke or to increase CO to dangerously high levels as load changes.

Compare this minimum value of excess oxygen to the expected value provided by the boiler manufacturer. If the minimum level you found is substantially higher then the vendor's, burner adjustments probably can improve fuel and air mixing, thereby allowing operation with less air.

8. Establish the excess oxygen (buffer zone) margin above the minimum value required for fuel variations, load changes, and atmospheric conditions. Add this to the minimum value and reset burner controls to operate automatically at the higher level-the lowest practical setting at the particular firing rate.

9. Repeat steps 1-8 for each firing rate being tested. For some control systems, it is not possible to establish the optimum excess-oxygen level at each firing rate. The reason is that control adjustments at one firing rate may also affect conditions at other firing rates. In such cases, choose the settings that give the best performance over a wide range of firing rates. A trial-and-error approach, one involving repeated tests, may be necessary.

Many experts agree that it generally is best not to make any adjustments to your control system in the lower control range of your boiler without being very careful. Air flow requirements at low-fire conditions usually are dictated by flame ignition characteristics and stability rather than by efficiency. Air/fuel ratios at low loads and at or near light off conditions are very sensitive and any changes may jeopardize safe light-off characteristics. If boiler load requirements force a boiler to operate at low loads much of the time, check with the boiler manufacturer's service group or a qualified combustion consultant before establishing excess-oxygen levels.

10. Verify that the new settings can accommodate the sudden load changes that may occur in daily operation without adverse affects. Do this by increasing and decreasing the load rapidly while observing the flame and stack. If you detect undesirable conditions, reset the combustion controls to provide a slightly higher level of excess oxygen at the affected firing rates. Next verify these new settings in a similar fashion. Then make sure that the final control settings are recorded at steady-state operating conditions for future reference.

Repeat these checks at frequent intervals until it becomes obvious that the boiler is not having problems that, on occasion, cause it to exceed smoke or CO limits or that control, burner or fuel system problems are not causing unsafe conditions to develop. It is easy to hide such problems by making high excess oxygen adjustments. Trying to optimize performance will cause these problems to reemerge.

When an alternative fuel is burned, perform these same tests and adjustments for the second fuel. It is not always possible to achieve optimum excess oxygen levels for both fuels at all firing rates. Based on information gained from the tune up

procedure, a judgment can be made as to the best conditions which are practical.

Evaluation of the New Low O_2 Settings

If energy gains are to be realized, the new low O_2 settings must be realistic and they must be maintained. Pay extra attention to furnace and flame patterns for the first month or two following implementation of the new adjustments. Thoroughly inspect the boiler during the next shutdown. To assure high boiler efficiency, periodically make performance evaluations and compare with the results obtained during the test program.

Review of the fine tuning process

It is sometimes possible during the optimization program to lower the CO or smoke limit, to achieve even lower excess air levels achieving greater efficiency gains. If the burner and fuel system is not functioning properly your best efforts at lowering excess air may be wasted. The approach to this procedure is to insure that everything is in conformance with the manufacturer's recommendations and then conduct organized "trial-and-error" (**Table 9.2**) adjustments in such a way that meaningful comparisons can be made. Items that may result in lower minimum excess O_2 levels include:

- Burner register settings

- Oil gun tip position

- Diffuser position

- Fuel oil temperature

- Fuel oil atomizing pressure

- Coal spreader adjustments

- Coal particle size

The effect of these adjustments on minimum O_2 are variable from boiler to boiler and difficult to predict.

The principal method used for improving boiler efficiency involves operating the boiler at the lowest practical excess O_2 level with an adequate margin for variations caused by fuel property changes, changes in ambient conditions, and the repeatability and response characteristics of the combustion control system.

Step-by-step Boiler Adjustment Procedure for Low Excess Air Operation.
1. Put the control system in manual control and bring the boiler to the test firing rate.
2. After stabilizing, observe flame conditions and take a complete set of readings.
3. Raise excess O_2 1-2%, allowing time to stabilize and take readings.
4. Reduce excess O_2 in small steps while observing stack and flame conditions. Allow the unit to stabilize following each change and record data.
5. Continue to reduce excess air until a minimum excess O_2 condition is reached.
6. Plot CO or Opacity versus O_2.
7. Compare the minimum excess O_2 value to the expected value provided by the boiler manufacturer. High excess O_2 levels should be investigated.
8. Establish the margin in excess O_2 above the minimum and reset the burner controls to maintain this level. This is the operating "Buffer Zone" which is based on an estimation of the amount of repeatability in the control system and the affects of other influences like temperatures and pressures.
9. Repeat steps 1-8 for each firing rate to be considered. Some compromise in optimum O_2 settings may be necessary since control adjustments at one firing rate may affect conditions at other firing rates if there is no means to characterize the air/fuel ratio.
10. After these adjustments have been completed, verify the operation of these settings by making rapid load pick-ups and drops. If undesirable conditions are encountered, reset controls.

Table 9.2 Boiler Tune-Up Procedure.

Chapter 10

Over 50 Ways To Improve Efficiency

Improving Efficiency

This chapter introduces over 50 ways to reduce the costs of operating boilers and distribution systems. These options are the result of tests and investigations at over a thousand commercial, institutional, industrial and utility plants around the world.

Although more than 50 efficiency improvement options are listed in this chapter, only a few will be appropriate at any one plant. The full impact of each proposed change must be understood; including (a) cost to benefit of the proposed change, (b) safety and (c) the impact of the change to the existing system and operations of the plant.

In most cases a complete analysis of a plant's operation is necessary to identify the causes of wasted energy and to identify the most cost effective remedy. This should involve professionals equipped with a good assortment of instruments for boiler diagnostics.

Smaller boilers are a lot more difficult to optimize than larger units. Smaller boilers are, because of "first cost" considerations, in a very competitive market. They are much less efficient, offering good opportunities for savings.

In larger boilers the original design often includes economizers, good burners and controls and oxygen trim systems. This is not always the case however, and there are many plants which could use this equipment especially in an era of high fuel costs and uncertain supplies. Often, low cost operational improvements and minor repair and adjustments are very cost effective at bigger plants.

This chapter has been written with two goals in mind. First it serves as a guide to assist in identifying energy wasting problems for any size boiler. Second, it can serve as a valuable educational tool, showing the logic of how to approach this complex problem.

Over 50 Ways to Improve Boiler Efficiency

1. **Manage Boiler Loads.**

      Cost: Low

      Efficiency Increase Potential: 2% to 50% +

      Description: Having specific efficiency information for each boiler, they can be matched to loads for most efficient utilization. Also, by using smaller boilers, which maintain high efficiency during very light load periods, losses associated with larger boilers can be eliminated. The general answer is to use the most efficient boiler combinations possible to match system demands.

Most boilers do not have a straight line efficiency and operate at different efficiencies depending on their firing rate and this must be considered when programming their operation **(Fig 10.1)**.

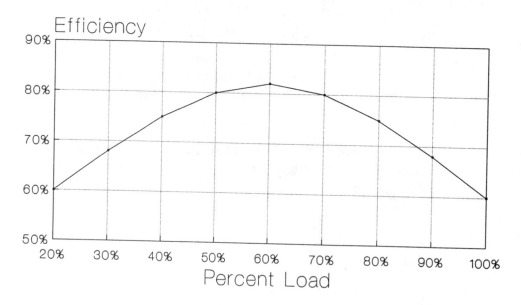

Figure 10.1. Boilers have different efficiencies at different firing rates.

122

Over 50 Ways to Improve Boiler Efficiency

Some boiler plants are over sized for their current loads, or small boilers are not available to handle light loads of summer or partial plant production schedules. **Figure 10.2** shows a typical heating system load profile where a boiler is at full load for only 20% of the time and operates at less than half load for more than 55% of the year. Boilers typically have very poor efficiency at low loads so this type of operation can be quite inefficient.

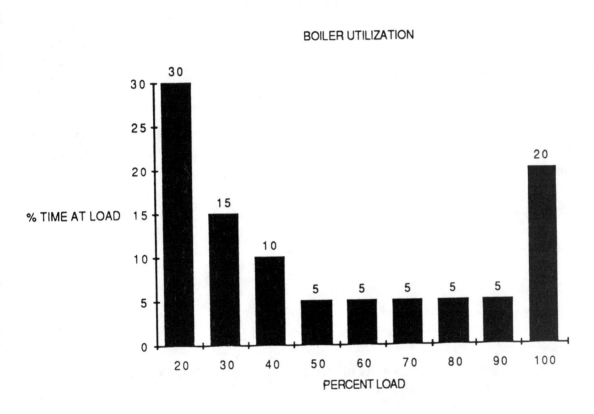

Figure 10.2, Actual boiler utilization for institutional type application.

The term "Annual Fuel Utilization Efficiency" **Fig 10.3** is used to describe the effect of this problem more accurately. Often boilers with measured efficiencies of over 80% have been found to be operating at less than 60% efficiency on an annual basis.

There are many solutions to this problem, which usually starts with the original design where large safety factors were employed to insure against an inadequate steam supply of steam on the coldest and windiest day of the year. Insulation improvements, energy conservation measures for the buildings and mild weather most of the time can really put these over-designed, high capacity systems in the energy wasting mode. This is caused by the boiler cycling on and off continually because of a low load factor. Each time the boiler shuts down or lights off again, cold air is blown through it to purge out any possible accumulations of unburned fuel as a safety precaution. This plus its normal heat loss to the environment can add up to a significant percent of its operating fuel demand.

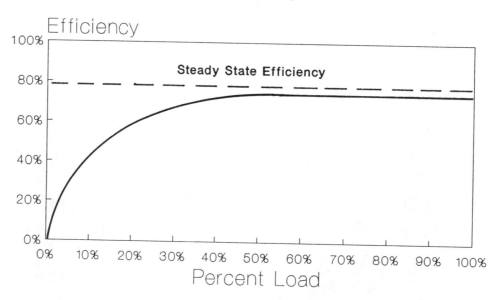

Seasonal Efficiency Partial Loading Curve.

Figure 10.3, Seasonal efficiency falls off with load, a disadvantage of oversized units.

The answer to managing boilers efficiently is simply stated: "You have to know the efficiency of your boilers at all firing rates and you have to take into account these standby losses."

With proper information, you can know about the efficiency of your plant under all operating conditions which will allow you to make effective choices about the most efficient way to operate them and their systems as well as have a basis for intelligent investments to improve efficiency.

Advantages/Disadvantages: Knowing the losses incurred at all levels of operation and having a plan to minimize them is a good management practice. Although it may be take extra effort to get the real time performance data required for true operating efficiencies of boiler systems, it must be done if energy dollars are to be saved. In some cases a continuous computerized monitoring of the system may be required. Without a proper analysis of annual fuel utilization efficiency, it may be difficult to select the most effective energy conservation options.

2. The 2M system: Measure and Manage.

Cost: Moderate

Potential savings: large

Description: There is virtually unlimited access to energy by its users, just throw a switch or open a valve and energy is available for immediate use. The question is,"Are the switches thrown and the valves managed responsibly?"

Large savings have been recorded just from monitoring energy use, no black boxes were involved. Flow meters and other instruments were installed and monitored. The most remarkable discoveries were made. Pumps and motors ran when they weren't needed, steam to process equipment wasn't turned off when the shift left for the day and boilers were being fired with no load on them, etc. This is a partial listing, but the lesson is clear, without accountability for energy use, without incentives to save utility dollars, waste and indifference will exist.

The key to discovering this type of loss is having instruments installed and monitored. It becomes very clear in a short period, when and where energy is being squandered. Without proper instrumentation and monitoring you will be running blind. The Gas and Electric Utility Companies don't give away energy, they use meters and monitor them closely. It's a good practice to follow when you consider the money spent purchasing their energy.

Advantages/disadvantages: Instruments are often expensive and in themselves do not account for any energy savings, they are however, the only way to find problems. Manpower that may be needed elsewhere must be assigned to collect and interpret the data from field measurements. It's just good management to know what is happening with energy, assigning responsibility for its use and controlling it in a sensible way.

3. Monitor and Manage With a Microprocessor Based System.

Cost: Moderate

Potential savings: large

Description: There is a definite reduction in manpower requirements for monitoring energy use with modern microprocessor based systems. The microprocessor is a powerful new tool that can take the pulse of your plant, detect problems and calculate efficiency every few seconds.

The computerized system **(Fig 10.4)** can store a tremendous amount of data and provide clear and simple reports and graphics. With "Intelligent Programs" the computer can train your personnel and help them to diagnose problems very quickly. With a telephone connection, a plant can be monitored by experts thousands of miles away.

The computer is truly the answer to tracking plant performance. Let's face it, a boiler is only a means of getting heat and power so the facility can accomplish its mission. If steam could be brought into a plant like electricity, many managers would be very happy because a great burden and liability would be lifted from their shoulders. Until this is possible, the next best thing to aid trouble free and efficient boiler plant operation is the use of the computer as a tool to help keep the operation of the plant smooth and trouble free as well as identify where energy (dollars) are being wasted.

Many boilers are unattended or are operated by a staff with many other jobs. With a computerized system, the boiler is monitored automatically which can mean a large savings in manpower and energy.

Advantages/Disadvantages: High cost and increased complexity are drawbacks. Properly used and maintained, a microprocessor can save a great deal of money by reducing labor costs, identifying fuel wasting problems, diagnosing plant and system problems and eliminating many service calls.

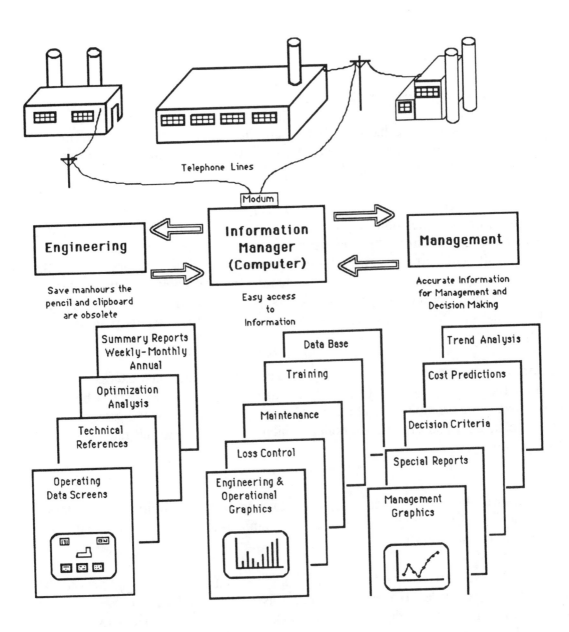

Figure 10.4. Quality information about plant operations is available for engineering, operation and management needs when microprocessor based systems are used.

4. Lead-lag boiler controls.

Cost: Moderate

Potential savings: Large

Description: The basic problem here lies in the operation of more boilers than is actually needed to meet current heat demands. The penalty for this type of operation is a 100% loss from the boiler that isn't needed.

If boilers are cycling on and off continuously, 15% to 20% or even greater losses can be expected. For example, boilers go through a light off purge cycle where cold air is used to clear any fuel vapors from the combustion zone each time the burner shuts off. Also, on the next light-off, cold air is again used to purge any possible vapors or fuel that may have accumulated during the off period. This plus the general cooling of the boiler during the off period requires all of the energy from the first several minutes of operation to reheat the boiler to its normal operating temperature. This energy is not reaching its point of use and is being wasted.

Burners become inefficient at low loads because of the additional excess air requirements for good air-fuel mixing. So the opportunity for loss is obvious when several boilers are operating at low load. One example is to eliminate "hot standby" boilers which essentially operate at zero percent efficiency and to use flash boilers which can come up to full rated capacity in 5 minutes instead.

The answer to this is to use a control system to stage boiler operation to avoid these losses. This is done with lead/lag controls that can be programmed to match and anticipate system demands avoiding low load operation.

Advantages/Disadvantages: There is the initial cost of installation and then follow up maintenance to consider. However these controls can start up boilers to have them ready to come on line when they are needed. They can work out problems of shared loads and take some of the workload from operating personnel. If they are not set up properly they will become a nuisance and fall into disuse.

Over 50 Ways to Improve Boiler Efficiency

5. Improved Instrumentation.

Cost: Low cost

Potential savings: Moderate to 25% (plus)

Description: There is a difference between "operating" and "managing" a boiler. To manage a boiler you need more information than is generally available with standard instrumentation which only gives information about "operating" conditions. Fuel flow, steam flow, temperatures and make up water use all play an important part in knowing what is happening and tracking trends. A boiler with only 5% utilization or with 300% excess air or a stack temperature of 750° F will not be efficient but it will produce steam and for all practical purposes be operating satisfactory. Likewise, a sudden increase in make up water use may indicate a large loss of energy in the distribution system. Good instrumentation is needed to identify these efficiency related problems.

Plants should have the following instruments:

    a. Stack thermometers.
    b. Fuel meters for each boiler.
    c. Make up feed water meters.
    d. Oxygen analyzers.
    e. Run time recorder.
    f. Energy output metering, steam flow, BTU out, etc.
    g. Return condensate thermometers.

The ideal scene would to employ a modern computerized monitoring system to continuously check on performance. If this can not be done, then this minimum instrumentation should be considered.

Performance should be tracked to meet fuel reduction objectives. A simple formula can be used if proper instrumentation exists:

$$\text{EFFICIENCY \%} = \frac{\text{BTU OUTPUT}}{\text{BTU INPUT}} \times 100$$

There are many portable boiler efficiency analyzers available which can be used to help assess the efficiency of your boilers. These range from simple wet chemical analyzers to electronic flue gas analyzers that can measure excess air, stack emissions and stack (heat loss) efficiency at the touch of a button. These instruments measure stack losses but do not give information about overall plant performance. Because the instrumentation at most plants does not provide continuous or complete information, there are a lot of blank spaces and assumptions about actual plant performance.

Over 50 Ways to Improve Boiler Efficiency

Advantages/Disadvantages: Instruments add extra cost to plants and in many cases they have not been installed or have been neglected. They must be kept calibrated and someone must record and analyze the data and then take positive steps to correct energy wasting conditions. This takes time, effort, money and skill. However, there may be a large pay off in knowing what is happening and discovering where the fuel wasting conditions are.

6. Fuel Selection.

Cost: Low cost, basically analysis and decision making
High cost, if major equipment change-out is involved

Potential savings: Large

Description: The bottom line is producing steam for heat and power at the lowest cost possible. At first glance you might think that the "best buy" would be using the fuel with the most BTUs per dollar. This type of ranking would show that wood, coal and heavy oil would be the best buy.

There is more to know about the situation than BTUs per dollar. There is also the conversion efficiency **(Table 10.1)** and related problems with transportation and pollution to consider. Knowing the cost advantages and problems associated with different fuels can lead to informed decisions.

Fuel Oil No. 6	85.95%	Lignite	80.86%
Fuel Oil No. 2	84.95%	Bituminous	87.42%
Natural Gas	81.30%	Anthracite	89.36%
Propane	83.45%	Methane	81.20%
Wood	71.11%	Hydrogen	77.47%

Table 10.1. Energy conversion efficiencies with different fuels.

Analyze the real cost of steam:

Fuel Type	Cost Per Unit	Cost Per Million BTUs	Expected Efficiency	Cost Per million BTUs or 1,000 Lb of Steam
Wood		$1.00	70%	$1.43
Coal	$40/Ton	$1.67	80%	$2.08
		$2.50		$3.13
Natural Gas	$0.287/Therm	$2.87	80%	$3.58
	$0.475/Therm	$4.75		$5.94
No.6 Fuel Oil	$0.50/gal	$3.26	85%	$3.85
	$0.65/gal	$4.24		$5.00
No. 2 Fuel Oil	$0.50	$3.57	84%	$4.25
	$0.65	$4.62		$5.53
Electricity	$0.04/KWH	$11.72	97%	$12.08
	$0.08/KWH	$23.43		$24.16

Table 10.2 The price of steam per million BTUs can vary to a large degree when considering basic fuel cost and conversion efficiency. Unfortunately, this analysis is only the tip of the iceberg and additional study may be needed. Each type of fuel requires different transportation to the plant and some like heavy oil, coal and wood require additional conditioning before they can be burned.

Over 50 Ways to Improve Boiler Efficiency

When confronted with the choice of switching from natural gas to residual oil, a more detailed study should be made. Notice that there are many hidden costs in the conversion to heavy oil that should be considered. This is a plus-point minus-point evaluation example. The (+) in front of an item means that it generally will be a benefit and a (-) indicates that there are drawbacks or additional costs involved.

Conversion To Oil	
+	Higher Combustion Efficiency
-	Cost to preheat oil
-	Cost to atomize oil
-	Soot blowing
-	Tank farm heating
-	Pumping costs
-	Surface temperature rise between soot-blowing
-	Long term stack temperature rise
-	Investment cost to maintain inventory
-	Environmental costs
-	Maintenance of boilers, tanks, atomizing systems, soot blowers, piping, pumps, burners and fuel oil heaters.

Figure 10.5, Plus point minus point evaluation table for conversion from natural gas to residual fuel oil.

Different costs will be appropriate for each of these items at individual plants. The important thing is to be aware of the fact that a complete analysis is needed. A simple overlooked fact is that you pay for natural gas after it is used and fuel oil must be purchased and stored for many months or years before it is used. This ties up capital which can be used for other purposes. Another real problem is the cost of environmental compliance in some regions as well as the uncertainty of future environmental requirements for storage tanks and underground piping.

Although the use of electricity to generate steam would seem to be an unwise choice, the whole picture should be considered. Distribution systems can waste a very large percentage of heat, especially if they are lightly loaded. It may be more cost effective to use electric boilers at those locations with only intermittent or light steam use, especially if long or inefficient steam and condensate runs are involved.

Advantages/Disadvantages: A complete analysis could be time consuming, especially if the plant would have to be reconfigured to burn another type of fuel, but on the other hand going through the trouble could be quite profitable.

7. Cogeneration.

Cost: High

Potential Savings: High

Description: Because of the high price of electricity in some sections of the country, it is very cost effective to use a cogeneration plant to provide steam and power while producing electricity. A utility plant usually wastes about 63% of the input energy and another 5-8% can be lost in transmission and transformers.

Some cogeneration facilities are over 80% efficient. This plus the high cost of electricity puts a cogeneration facility at a good economic advantage. Some calculations have shown that to match the bottom line (in dollars) of a cogeneration facility, a boiler would have to have an efficiency of 120%-130% or higher.

Peak shaving is another consideration, a large part of many electric utility bills is for "demand" charges which amounts to a payment for the peak demand of electricity and not actual energy use. Often the peak demand for electricity occurs at the same time as the demand for other types of energy such as heat and cooling, so if the on site generation of electricity can reduce electrical demand and provide other forms of energy at the same time, large savings are possible.

The key to effective cogeneration is to have both electrical and heat demands occur at the same time.

Advantages/Disadvantages: This option may take an extensive analysis but should not be overlooked as a way to improve the bottom line of your operation. High cost, load matching, utility intertie, red tape and pollution control are a few of the challenges that must be faced with cogeneration. It is a very good way to reduce steam and electrical power costs and may be well worth the trouble.

8. Boiler Tune-Up.

Cost: Low

Potential Savings: 2% to 20%

Description: A good tune-up, using precision test equipment, can detect and correct excess air losses, smoking, unburned fuel losses, sooting, fire side fouling and high stack temperatures.

By using real time instruments to measure flue gas oxygen, carbon monoxide, carbon dioxide, combustibles and temperatures, the current state of boiler operation can be diagnosed. These findings can be used to restore the boiler to its normal efficient operating condition (**Figure 10.6**). This allows the control system and burner to be adjusted and repaired for optimum performance with immediate feedback of results for gas fuels (**Figure 10.7**) and oil (**Figure 10.8**).

When a tune-up is completed there should be a good record of a boiler's excess air and efficiency across the load range for managing boiler operations.

After the initial tune-up, recheck results until assured that the boiler can stay tuned-up for a reasonable amount of time. Tune-ups may be needed more than twice a year. Periodic checks will tell the story; often boiler efficiency will fall by over 5% within 6 months or sooner.

Advantages/Disadvantages: If you intend to manage a boiler plant and control losses, regular tune-ups and combustion efficiency checks are necessary. It takes skilled manpower to run these checks and this is an additional expense so judgment should be applied to establish the benefit to cost ratio of such a program.

(**Very Important:** All energy conservation options should be evaluated from the tuned-up condition to eliminate false estimates of the benefits of retrofit options.)

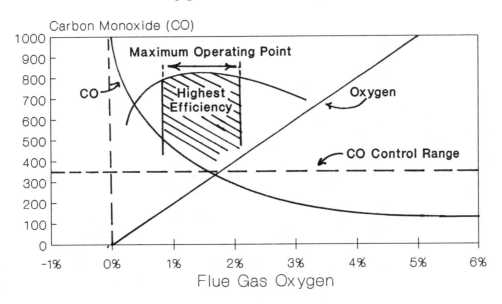

Figure 10.6, Carbon monoxide vs. oxygen relationship showing CO control range and operating range for highest efficiency.

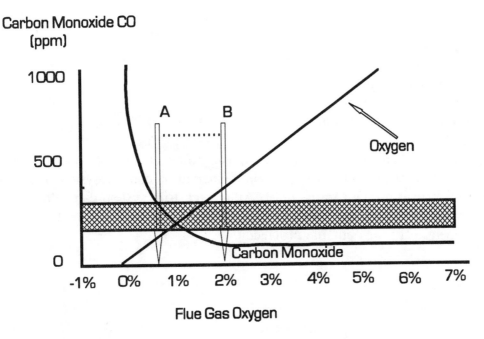

Figure 10.7, Carbon monoxide limit vs. oxygen relationship for adjusting controls for optimum efficiency. Point (A) is minimum oxygen control point, (B) is set point based on buffer zone for various system errors (dashed line).

135

Over 50 Ways to Improve Boiler Efficiency

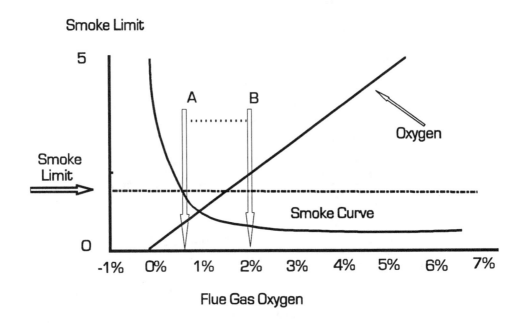

Figure 10.8, Smoke limit vs. oxygen relationship for adjusting controls for optimum efficiency. Point (A) is minimum oxygen control point, (B) is set point based on buffer zone for various system errors (dashed line).

9. Oxygen Trim.

Cost: Moderate to low.

Potential Savings: Moderate

Description: There are many factors that can influence and introduce errors into the air/fuel ratio such as fuel property changes, air temperature, control system response, fuel pressure and burner performance **(Table 10.3)**. An oxygen trim system can automatically and continuously compensate for the variables in the combustion process, insuring that the boiler is programmed to operate at near optimum efficiency.

Over 50 Ways to Improve Boiler Efficiency

Sources of Errors in Control Systems

Fuel	Combustion Air	Exhaust	Controls
Pressure	Temperature	Fouling	Fuel positioning accuracy
Temperature	Humidity	Stack effect	Air positioning accuracy
Viscosity	Pressure		Control system accuracy
BTU Content	Fan performance		Control alignment
Atomizing performance	Dampers		Control synchronization

Table 10.3. The sources of errors in control systems.

An oxygen trim system can compensate for the system errors, allowing a lower excess air level to be maintained on an automatic basis (**Fig 10.9**). However, many systems have not taken full advantage of this potential and are maintaining a higher than necessary set point for excess oxygen. The O_2 trim system can also be used as a tool to diagnose and control system problems.

Oxygen Trim

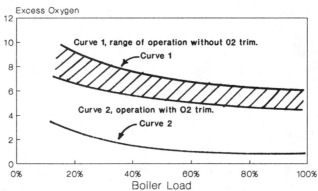

Figure 10.9. An oxygen trim system compensates for many control system errors.

Advantages/Disadvantages: Without oxygen trim, high levels of excess oxygen are necessary to compensate for many sources of errors in the combustion process; this excess oxygen can waste a great deal of energy. Oxygen trim systems do not seek the most efficient operating level. They are set to a level of excess oxygen established by tests using a carbon monoxide or smoke limit. Some latitude must be given in making this operational setting, so it does not end up at the most efficient fuel/air ratio because it includes a buffer zone too. Also, the oxygen trim system must be maintained in good operating condition and calibrated, adding to the plant maintenance challenge.

Over 50 Ways to Improve Boiler Efficiency

10. **Carbon Monoxide Trim.**

      Cost: High

      Potential Savings: Moderate

      Description: A major problem with the oxygen trim system is that it can not seek out the most efficient operating point and it must operate at fixed oxygen set points. For this reason a fuel wasting buffer zone is programmed into the operation of oxygen trim systems. A carbon monoxide measurement and trimming system is designed to seek out the most efficient operating level on a continuous basis, insuring the most efficient operation possible **(Fig 10.10)**.

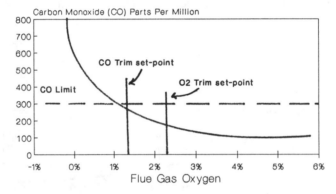

      Figure 10.10. A carbon monoxide trim system continuously seeks the optimum air/fuel ratio. In contrast, the O_2 trim system must be set to a fixed value and does not sense poor combustion, combustibles or formation of carbon monoxide.

      Advantages/Disadvantages: Carbon monoxide trim is more expensive and requires more maintenance. In the past it has had a poor reliability record in the harsh combustion flue gas environment, but under the right conditions it can save a great deal of money. These are sensitive instruments that require special attention in maintenance and calibration. The CO trim system delivers its best performance and payback when all other parts of the combustion process are working perfectly.

Over 50 Ways to Improve Boiler Efficiency

11. Control System Linearity.

Cost: Low

Potential Savings: High

Description: When control systems have not been designed or calibrated for linearity they do not respond properly. If you could compare a boiler's control system to an automobile's accelerator, linearity error is like just resting your foot on the gas pedal and having it accelerate to full speed with very little foot motion. In another case, having your car not respond to large change in position of the gas pedal and only getting a response on the last sixteenth of an inch of travel **(Figure 10.11).**

Other conditions poor control system linearity may cause, are that a boiler may only operate between certain restricted loads, like from 40% to 65% or from 80% to 100%. This is often a hidden error which makes smooth boiler control almost impossible and it takes a good analysis to uncover.

This is one of the many good reasons for equipping boilers with good instrumentation, including fuel meters. This condition will cause a boiler to cycle on and off or it may cause a boiler to operate below rated capacity making it necessary for additional boilers to carry the load. Both conditions waste fuel and makes it very difficult to do a good tune-up.

When a boiler is first put into service, control system response should be checked to insure that its operation is linear. Many boilers have this hidden problem which interferes with efficient operation and makes life very frustrating for operators.

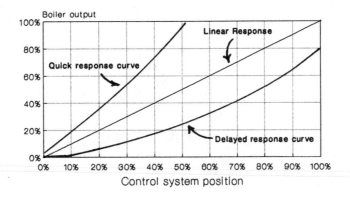

Figure 10.11. Control system linearity. Non-linear response can cause control problems and inefficient operation.

139

Over 50 Ways to Improve Boiler Efficiency

Advantages/Disadvantages: It takes time and additional expense to insure that a boiler's firing system is linearized and operating in its rated operating range. It's one of those things taken for granted when equipment is new. Unless this problem is resolved, it may be difficult to optimize the performance of a boiler to the fullest extent.

12. Characterizable Fuel Valve.

Cost: Low

Potential Savings: 2% to 12%

Description: A characterizable fuel valve **(Fig 10.12)** has a series of adjustments that are used to match the air/fuel ratios across the load range. Without this type of valve, a precise tune-up is almost impossible. It also serves as a valuable tool to correct mechanical problems that can develop in the burner and control systems. One valve is needed for each fuel.

The reason a characterizable fuel valve can be so useful becomes apparent when the fuel delivery curve and air delivery curves are compared **(Fig 10.13)**. Often they do not match and many control systems have no way to adjust one to the other in a precise way without sacrificing excess air and lost energy.

FUEL VALVE

CHACTERIZABLE
CAM

Figure 10.12. Characterizable Fuel Oil Valve.

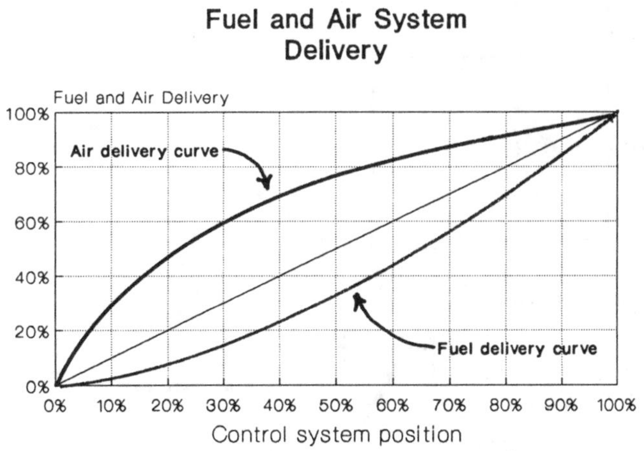

Fuel and Air System Delivery

Figure 10.13. Different delivery rates from air and fuel systems.

Advantages/Disadvantages: This is a low cost option and one of the best investments you can make for boiler optimization and troubleshooting. There should be a characterizable fuel valve for each fuel fired if the control system has no other means to balance fuel flow with air flow.

13. **Over Fire Draft Control.**

Cost: Low

Potential Savings: Moderate (2% to 10%)

Description: One of the first and simplest devices used to control excess air was with over fire draft control. Because of the change in draft on cold and hot days and the lack of precise draft control experienced with barometric dampers, over fire draft control systems are used to maintain a   constant negative pressure in the fire box. This establishes predictable conditions for air fuel ratio control. Also, when the boiler shuts down, the over fire draft control system damper closes, preventing heat from escaping up the stack **(Fig 10.14)**.

Advantages/Disadvantages: The over fire draft control is a simple, low cost ($1,500) device and much cheaper than oxygen trim.

Over 50 Ways to Improve Boiler Efficiency

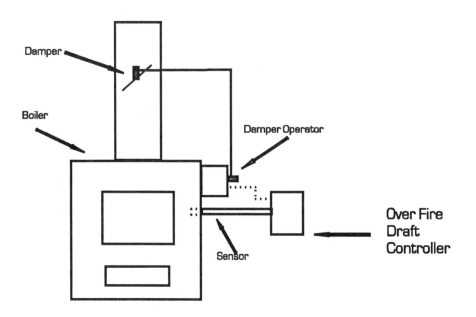

Figure 10.14. Over Fire Draft Control system.

14. Control Exhaust Draft Conditions

Cost: Low

Savings: Moderate to low

Description: In many cases the size of stacks and exhaust systems is excessive allowing unrestricted and often excessive negative draft draw to exist. Oversized exhaust systems offer little resistance to the varying conditions of stack "draw" and cause excess air to be pulled into the burner as well as aid the unrestricted escape of large volumes of high temperature exhaust gasses. Years ago, slots were cut in the stack and metal dampers inserted, but these often were unsatisfactory. A modern adjustable design has been developed, a special exhaust gas buffer that has shown good results **(Fig 10.15)**.

Advantages/disadvantages: This inexpensive device can slow down the escape of hot gases up the stack. It can also be adjusted to a specific location.

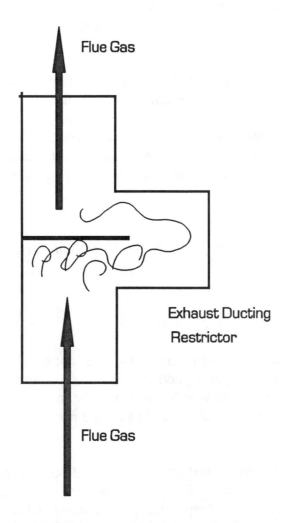

Flue Gas

Exhaust Ducting

Restrictor

Flue Gas

Figure 10.15. A fixed damper buffers high velocity escape of flue gas without limiting cross sectional area of stack.

15. **Cross Limiting Controls.**

    Cost: Moderate

    Potential Savings: Moderate

    Description: Fuel delivery responds almost immediately to control system positioning whereas it takes longer for the air to respond, so on upward swings you can get a black smoking condition until the air catches up with the fuel.

This smoking on load changes is a waste of fuel and the soot deposit it leaves behind fouls the heat exchange surfaces causing higher stack temperatures. This is one of the reasons that

143

pneumatic and electronic controls are used. They can be designed for this cross limiting feature which prevents smoking. If a boiler is smoking on load swings, then it is changing load too fast, and cross limiting is needed.

Cross limiting systems simply do not allow the fuel valve positioning signal to be greater than the actual measured air flow requirement, thus eliminating smoking on load changes.

ar   **Advantages/disadvantages:** It may be expensive to install a completely new control system just to have cross limiting. However, if this condition is not corrected, the price in unburned combustibles and higher stack temperatures will surely be taking an undesirable share of your fuel budget.

16. Improve Control System Accuracy With Strong Precision Parts.

Cost: Low

Potential savings: Moderate (2% to 12%)

Description: If your control system is flimsy and does not have strongly built precision parts, the chances are that you can't get a good tune-up or repeatability. If control system linkage has any play at all, precise excess-air levels cannot be maintained. For example, only 1/100 of an inch play in control system linkage can cause the need for a 5% excess air buffer zone.

If low excess air levels are to be maintained and smoking and soot formation prevented, the linkage should be precise with no play. Sometimes pins are lost and threaded bolts are substituted instead and over time, these sharp threads cause elongated holes and a tremendous amount of play. Also, non-precision parts are used to substitute for lost or damaged parts.

Control linkages also bind, warp, shear pins and otherwise lose their precision over the years and the only solution is to restore it to a precision system.

Advantages/Disadvantages: You may have a burner and control system that can't be adjusted to maintain low excess air levels and high efficiency. If so, some custom rebuilding may be necessary. It's a cheap enough solution but you may have to be innovative and use aircraft grade precision parts and such.

Over 50 Ways to Improve Boiler Efficiency

17. Stack Dampers.

Cost: Low

Potential Savings: Moderate (5% to 20%)

Description: When a boiler isn't firing, a great deal of energy can be lost up the stack due to the chimney effect when the lighter hot air formed in the boiler internals and in the hot stack pull cold air in through the boiler. This situation is most common on atmospheric burner units, but may be present on some older equipment too.

A damper automatically closes off the stack when the boiler isn't firing, holding the heat in the boiler (**Fig 10.16**).

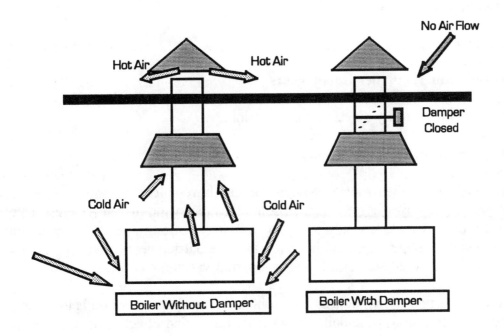

Figure 10.16, Stack dampers used to block heat losses.

Advantages/Disadvantages: Stack dampers have full safety approval and should be installed to curb this unnecessary loss.

18. Vent Caps on Stacks

Cost: low

Savings: Low to Moderate

Description: Something as simple as a vent cap on the stack is not an obvious candidate for fuel savings. A strong wind and a bent or damaged stack cover can produce some wild draft effects. These draft effects will change the air/fuel ratio and cause large swings towards excess air conditions and possibly cause smoking.

With the lack of good draft control, excess air adjustments must be made to compensate for this effect which can be excessive causing wasted fuel.

It has always been a mystery as to just how vent covers get bent and pushed over to one side, but they have been known to deflect air into the boiler or furnace in a good wind and cause serious problems and can even back smoke into a building.

19. Wind Deflector for Boiler Room Vents

Cost: Small, basically a design consideration.

Savings: Moderate

Description: Boiler rooms are required to have a certain square footage of free ventilation access per boiler horsepower. In some locations wind driven air pressure can cause draft control problems. One has only to try to walk into a strong wind to grasp the magnitude of the force involved. When the wind is driven into the outside vents boiler room pressure can rise, when on the sheltered side of a building a partial vacuum can form.

This condition can cause burner adjustment problems. The simple solution is to be aware of the potential problem and protect the boiler room from this wind effect or set the excess air high enough to stay out of the smoking zone.

This may seem like a small thing, but when low excess air levels are to be achieved a condition like this can become a big challenge.

Over 50 Ways to Improve Boiler Efficiency

20. Low Excess Air Burner

Cost: Moderate to High

Potential Savings: Moderate to High

Description: It doesn't matter how good your controls are, or how fancy the trim system, the burner will ultimately control the level of excess air that can be achieved. Older burners may not have been designed with low excess air in mind or they may have worn out to a point where low excess air levels are no longer possible. The solution is to install a burner which has been designed for low excess air operation, or to repair or refurbish the one you have so that it is capable of dependable low excess air operation.

Advantages/Disadvantages: High cost and redesign of boilers to accommodate these new burners may be a problem. However, low excess air operation may not be possible without burner conversion.

21. Oxygen Enrichment

Cost: Varies

Savings: Moderate to small

Description: Substitution of oxygen for combustion air, which contains only 20.9% O_2, reduces the volume of heat absorbing nitrogen flowing through the combustion process and therefore reduces flue gas loss. Energy savings must be measured against the cost of oxygen and the cost involved in additional safety precautions.

The usual oxygen enriching practice in industrial heating is to increase the oxygen in the combustion air from its normal 21% to 25 or 30%. It is preferable to premix it with the combustion air, but lances are sometimes used. Economic justification of the use of oxygen is often marginal unless the user has oxygen facilities with spare capacity.

Advantages/Disadvantages: Good energy savings are possible but safety will be a paramount factor in mixing oxygen with fuel oil.

Over 50 Ways to Improve Boiler Efficiency

22. Oil-Water Emulsions

Cost: Moderate

Potential Savings: Low

Description: This is a process where very small amounts of water, usually less than 6% of the fuel, of micron sized droplets are evenly mixed with fuel to cause additional atomization action in the combustion zone.

The benefit from oil /water emulsions is limited, but is of great value in those plants where serious burner problems are causing smoke and excessive particulate emissions. When exposed to the intense heat in the fire box, the micron sized droplets of water expand rapidly causing the oil droplets to break into smaller particles, enhancing atomization. This reduces the smoke level allowing lower excess air and the elimination of the sooting of the heat exchange surfaces.

Advantages/disadvantages: This is an effective way to reduce particulate generation and change the smoke threshold so less excess air can be used. The equipment is expensive and the water used in this process escapes up the stack carrying away a high level of energy.

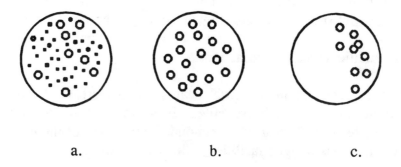

a.   b.   c.

Figure 10.17. Oil-water emulsions, micron sized water droplets are mixed with fuel. This figure shows typical atomized oil droplet with water mixed in fuel: (a) water droplets are too small, (b) water droplets uniform and the proper size and (c) non-uniform mixing of water droplets.

Over 50 Ways to Improve Boiler Efficiency

23. Fuel Oil Additives

Cost: Low

Potential Savings: Low (Studies have shown savings up to 2%)

Description: There are three classes of fuel additives:

Class I. Fuel handling, which includes sludge and gum inhibitors, detergents, metal deactivators, color stabilizers, pour-point depressants, anti-static and anti-icing compounds, and corrosion inhibitors.

Class II. Combustion additives for improved combustion and pollutant reduction; combustion improves to reduce smoke and particulates, carbon monoxide, hydrocarbons, nitrogen oxides and sulfur oxides.

Class III. Post-flame treatment additives which include soot removers, additives to control fireside corrosion and slag deposits, additives to enhance particulate collection and SO_x scavengers.

Advantages/Disadvantages: If the smoke point can be improved, lower excess air levels are possible. If there is less soot and fire side fouling, then lower stack temperatures can be maintained. Both conditions improve efficiency. Additives can help, studies have shown improvements up to 2%.

24. Replace Atmospheric Burners With Power Burners

Cost: Moderate

Potential Savings: Moderate

Description: An atmospheric type of burner has practically no excess air control and loses a lot of heat up the stack during the off cycle if a damper is not installed.

On the other hand, excess air and stack losses can be more easily controlled with a power burner which has electrically positioned control linkage, motor driven fans and automatic dampers.

Advantages/Disadvantages: Atmospheric type burners have no excess air control. Excess air levels have often been measured at extremely high levels. In addition, because they use air that is pulled into the combustion zone by the chimney effect there is no positive way to prevent cold from being sucked through the boiler when it is not firing, which wastes a lot of energy up the stack. A power burner has two clear advantages, it can be adjusted for excess air control and its dampers close during the off cycle, keeping heat in the boiler.

25. Flame Retention Head Type Burners

Cost: Moderate

Savings: Low (with new units)

Description: Without good air-fuel mixing, oil fired burners require higher amounts of excessair. Flame retention head burners, however, provide better mixing of oil and combustion air, thereby reducing excess-air requirements **(Fig 10.18)**. The application of this option depends to a great degree on existing excess-air levels. This option should be considered when replacing old equipment or in the design phase of mechanical systems.

U. S Department of Energy tests found an improvement in seasonal efficiency from 5.9 to 12.7 percent when high speed flame retention head burners were used on standard heating units.

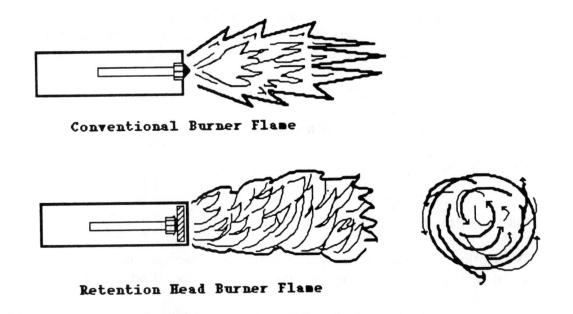

Conventional Burner Flame

Retention Head Burner Flame

Figure 10.18, Flame retention head burner.

Advantages/Disadvantages: This is a small boiler and furnace option. Flame retention head burners should be specified for any new purchases. They may be available as a retrofit option if the manufacturer has redesigned older models. It is a low cost item that can reduce excess air.

26. Multi-stage gas valves

Cost: Low

Saving: Moderate

Description: Gas fired furnaces and boilers having only one fuel delivery rate can be a disadvantage which wastes fuel because the only firing rate is full fire and off. Having the option of intermediate firing rates (high, low and off) can lower stack temperature and eliminate to some of the standby losses.

Advantages and Disadvantages: This option requires additional pressure regulating valves and possibly burner modifications to produce stepped firing rates.

27. Reduction of Fuel Firing Rate

Cost: Low to Moderate

Savings: Low to Moderate

Description: Conditions change, many burners have been found to be grossly oversized for their applications. Successful energy conservation actions, changes in load requirements and other factors may be producing far less demand for heat. If fuel firing rate can be reduced, then stack losses will decrease along with standby losses.

Advantages/Disadvantages: The cost of installing new burners and the cost of engineering and other fees for redesign of an existing system may not be justified. A study of the economics of the issue will show if it will pay. It makes no sense to drive a lot of heat into a boiler with high exhaust gas temperatures and then let it sit idle for long periods; this is wasteful. A longer firing time with lower losses makes more sense. Caution may be needed at low firing rates where the exhaust gas temperature is depressed below the acid dew point which can cause damage to the boiler, breechings and stack.

the economics of the issue will show if it will pay. It makes no sense to drive a lot of heat into a boiler with high exhaust gas temperatures and then let it sit idle for long periods; this is wasteful. A longer firing time with lower losses makes more sense. Caution may be needed at low firing rates where the exhaust gas temperature is depressed below the acid dew point which can cause damage to the boiler, breechings and stack.

28. Replace On/Off Controls with Modulating Controls

Cost: Moderate

Potential Savings: Moderate

Description: Replace on/off type of burner control with controls that can modulate and match load conditions, reducing higher stack temperatures and standby losses. On /off equipment must be set to match the highest expected demand and when the demand has been satisfied the equipment must shut off. Exhaust temperatures are lowered by matching the firing rate to actual demand. Also, the losses associated with purging the boiler before and after each firing cycle could be eliminated by having fewer on/off cycles. **Table 10.4** shows typical efficiencies and the enhancements of this type of load matching.

Control Type	Efficiency at % Load			
	25%	50%	75%	100%
On/Off	70.3	74.4	75.6	76.3
On/Off With flue damper	73.3	75.3	76	76.3
High/Low/Off	76.9	76.5	76.4	76.3
Modulating	76.9	77.7	77.2	76.3

Table 10.4. Control system performance comparison.

Advantages/Disadvantages: On/off controls are simpler, with a lower first cost and can be tuned up easily. However, they waste energy with higher stack temperatures and standby losses. The more complex and more expensive modulating controls can overcome this problem.

29. Convert to Air or Steam Atomizing Burners

Cost: Moderate

Potential Savings: Moderate 2%-8%

Description: Steam or air atomization allows the firing of a wide range of fuels and high turndown approaching 20:1 instead of 4:1, and good efficiency. Air and steam atomization produces an aerosol in which fine droplets are supported by an expanding cone of air/steam and there is less sensitivity to oil viscosity changes. It also allows more flexibility in shaping the flame to conform to furnace conditions.

Advantages/Disadvantages: The cost of installing and operating a compressed air or steam atomizing system may be a disadvantage. The steam used for atomization is lost up the stack and its energy must be considered. Offsetting this disadvantage of lost energy is improved performance, increased turn down ratio and fuel versatility; especially if fuel quality tends to vary.

30. Fuel Oil Viscosity Management

Cost: Low

Savings: Moderate

Description: The design of the atomization system is based on certain fuel oil properties, primarily viscosity. Viscosity is the relative ease or difficulty with which oil flows or is pumped. Viscosity affects the quality of atomization and the smoke point which determines the minimum excess-air possible. Because of this, the viscosity of fuel oil must be closely controlled at all times.

Since oil is obtained on the open market, its source of origin can be uncertain. Therefore, the properties of one batch will not likely be the same as those of another batch.

Oil viscosity is controlled by temperature. The oil temperature necessary at the atomizer depends on the particular fuel. Some heavy fuels are very fluid and require little or no heating. A viscosity of 180 to 200 Saybolt Seconds Universal (SSU) often gives best results for atomization and if the oil at ordinary temperature is of higher viscosity, it must be heated to a temperature which will reduce viscosity to this point.

Over 50 Ways to Improve Boiler Efficiency

The temperature control of the fuel oil heater is a key element in controlling atomization and smoke point for excess-air control. Fuel batches should be sampled to determine the temperature set point. If the heater does not maintain an accurate set point, close excess-air tolerances may not be possible. Automatic viscosity controllers will feed back information on viscosity to the heater and maintain a proper temperature.

Advantages/Disadvantages: In general, the lowest fuel temperature permitting good furnace conditions and low excess-air is most desirable. Too high a temperature may cause carbonization of fuel oil heaters and atomizer tips, sparking in the furnace and a tendency for the flame to become unsteady and blow off the tip.

Burner performance controls the minimum possible excess-air levels which can be held consistently. The burner depends for its performance on oil being at the proper viscosity. Poor viscosity (temperature) control can prevent high efficiency. A properly working and maintained fuel oil heater and good viscosity control either through regular tests on oil batches or with an automatic controller is the key to low excess-air.

31. Economizer

Cost: Moderate to High

Potential Savings: Moderate

Description: One of the most effective ways to recover energy from flue-gas is to use an economizer (**Fig 10.19**) to heat the feed water going into the boiler. At 100 PSIG the temperature of the steam and water in the boiler is about 334° F. The temperature approach of the flue gas to steam temperature is in the range of 50° F to 150° F, depending on firing rate. This can result in a stack temperature from 385° F to over 480° F providing a good source for heat recovery.

Hot flue gas contains a lot of wasted energy and it is very beneficial to put this energy back into the boiler with heat recovery equipment. The rule of thumb for reducing stack temperature is: for every 40°F the temperature is reduced, a 1% efficiency increase occurs. In this case lowering stack temperature by 150°F will result in an efficiency increase of approximately 3.75%.

Preheating the incoming feed water also improves efficiency. The usual temperature of incoming feed water is 212 F to 220 F or higher. This water can be heated in the economizer improving efficiency on the water system side.

Over 50 Ways to Improve Boiler Efficiency

The rule of thumb for water heating that applies here is: for every 11 degrees F you raise the feed water entering a boiler, the efficiency goes up by one percent. So, if you can take water from the deareator and heat it from 220°F to 277°F, a 5 percent (plus) fuel savings can be produced.

Advantages/Disadvantages: Economizers are fairly expensive and must be justified by a high level of boiler utilization. Once installed, they are usually trouble free and require little maintenance. If stack temperature or incoming economizer water temperature is too low, an acid formation problem may develop.

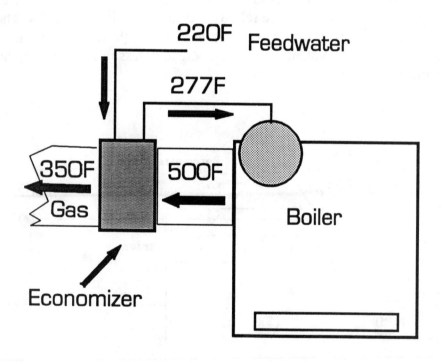

Figure 10.19, Economizer lowers stack temperature from 500°F to 350°F, raising feed water from 220°F to 277°F.

32. Heat Reclaimer

Cost: Moderate/Low

Potential Savings: Moderate

Description: The heat reclaimer **(Fig 10.19)** is similar to the economizer, except that it is used on smaller hydronic or hot water systems. It can also be used on boilers for heating process or domestic hot water.

A heat exchanger is installed in the stack with its own pump and regulating valve and is basically an additional heat exchange surface for the heating unit. The flue gas temperature can be controlled by regulating the flow of water through the heat reclaimer. Many of the older hydronic boilers have high stack temperatures and high excess air levels, making this option very attractive.

Advantages/Disadvantages: Design and installation costs on the smaller units may be hard to justify, otherwise they are relatively trouble free low cost units. The heat-reclaimer has two advantages, (a) it recovers waste heat and (b) it lowers excess air on atmospheric type boilers by cooling exhaust gasses creating less furnace draw in both the on and off condition.

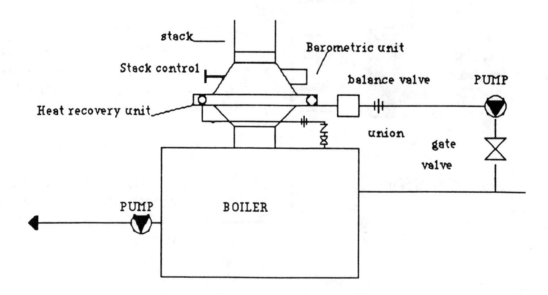

Figure 10.19, Heat recovery unit installed on a hydronic boiler.

Over 50 Ways to Improve Boiler Efficiency

33. Air Preheater

Cost: Moderate to High

Potential Savings: Moderate

Description: Another means to recover the wasted energy resource in the flue gas is the air preheater, which is used to heat the incoming combustion air. This is an excellent way to capture energy which would otherwise be lost and to put it back to work in the boiler. The rule of thumb which applies to this type of heater is that you can expect a one percent efficiency increase for every 40 degrees that you can decrease the net stack temperature (outlet stack temperature minus inlet combustion air temperature).

There are several types of air heaters:

The **tubular type** air heater circulates cooler combustion air around hot tubes which are heated by exhaust gases. Several novel but effective designs have emerged for this type of heater; glass tubes have been used for applications where corrosion is a problem. Also, teflon coated tubes are used if you wish to bring flue gas temperatures below the acid dew point. Heat pipes are also being introduced as an effective means to capture as much heat as possible.

Plate type heat recovery units also offer another excellent way to recover latent heat from the flue gas stream. They are good heat exchange devices and can be constructed of various materials including stainless steel, teflon coated steel and titanium to name a few.

The **rotating type** heater has segments which are heated in the exhaust stream and rotated into the cooler combustion air stream where they are cooled. This type of unit can be very large when used in utility plants. One drawback is that they must rotate continuously and good seals are needed to keep the exhaust gasses from crossing over to the combustion air stream.

Advantages and disadvantages: Air heaters are usually very large and the need for large supporting ducting systems to the burners is expensive. They are an excellent way to capture low grade heat. Also, using hot air for combustion raises the furnace temperature and there is a possibility that the refractory may be affected.

Over 50 Ways to Improve Boiler Efficiency

34. Turbulators

Cost: Low

Potential Savings: Low

Description: Turbulators are a very effective way to reduce stack temperatures and increase efficiency of fire tube boilers **(Figure 10.20)**. In a firetube boiler the hot gasses must pass through relatively long narrow tubes where heat is given up to the side walls causing the gasses to contract and slow down forming layers of colder gas along the heat exchange surfaces. This condition prevents good heat transfer, so turbulators are used break up this insulating film.

Turbulators are thin strips of metal that have been bent to break up the flow of hot gasses in the tubes and are designed for minimum obstruction of the passages. They are usually installed in the last pass. By using longer turbulators in the upper tubes of the last pass, hotter gasses are forced through lower tubes eliminating to some degree the stratification of hot gasses at the end of the boiler.

Advantages/Disadvantages: Turbulators may cause condensation of flue gasses when a boiler is first lit off. Generally, they are a good inexpensive way to lower flue gas temperature. They are not recommended for high particulate loading options such as coal and residual fuel oil.

35. Soot Blower

Cost: Moderate

Potential Savings: Moderate. The rule of thumb applies: every 40 degree decrease in net stack temperature saves about one percent in fuel consumption.

Description: Soot is an excellent insulator which can retard heat transfer to a great degree, so soot blowers are used to keep heat exchange surfaces clean to maintain lower stack temperatures and higher efficiencies. Also, there is more of a tendency to form soot with the heavier residual type fuels. On the other hand, soot blowing is an expensive proposition when you consider the loss of energy involved to just keep the tubes clean; it can cost from 50 to 200 thousand dollars a year for a single boiler, so it is important to manage soot blowing properly. Both too little and too much soot-blowing can waste significant amounts of energy. When low excess air levels are maintained, keeping heat exchange surfaces clean becomes a greater challenge because the burners are operating closer to their smoke point.

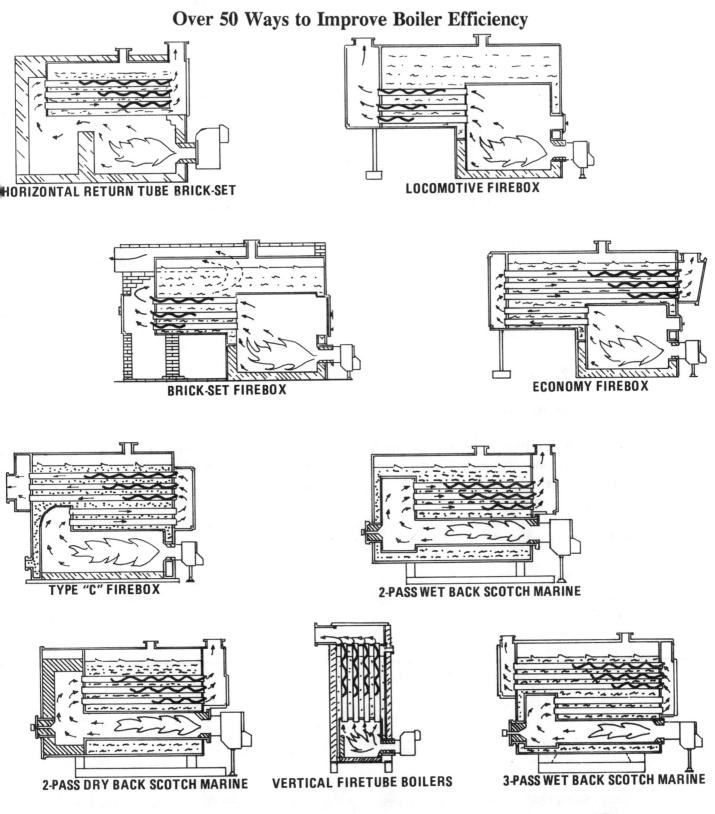

HORIZONTAL RETURN TUBE BRICK-SET

LOCOMOTIVE FIREBOX

BRICK-SET FIREBOX

ECONOMY FIREBOX

TYPE "C" FIREBOX

2-PASS WET BACK SCOTCH MARINE

2-PASS DRY BACK SCOTCH MARINE

VERTICAL FIRETUBE BOILERS

3-PASS WET BACK SCOTCH MARINE

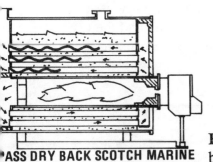

PASS DRY BACK SCOTCH MARINE

Figure 10.20, Various types of firetube boilers with turbulators installed.

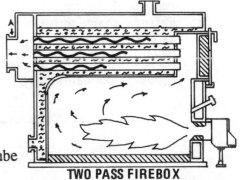

TWO PASS FIREBOX

Soot blowers are usually installed on water tube boilers, but they are also available for fire tube type units on special order for applications such as black liquor and other dirty fuel type service. Some plants have even had successful results with a large ship's fog horn to remove deposits with low frequency vibrations. They are less expensive to operate than the conventional units, but their performance has had mixed reviews.

Advantages/Disadvantages: Soot-blowers can do an effective job of keeping heat exchange surfaces clean and have been used for years. The cost of steam and compressed air to run soot-blowers can be very high so close control of their use is recommended.

36. Waste Heat Recovery Boilers

Cost: High

Savings: High

Description: There are many processes and applications where heat is wasted that could be captured in a waste heat recovery boiler especially where exhaust temperatures are 500°F and above. This type of system has been with us for a long time, but opportunities still exist to tap this virtually free source of steam and heat. The classic use of waste heat boilers is on reciprocating engines and gas turbines which have exhaust temperatures in the 1,000°F range.

Advantages/Disadvantages: First cost, additional maintenance and operational costs are involved in waste heat boiler planning. They can be an excellent source of free energy.

37. Blowdown Heat Recovery

Cost: Low to Moderate

Potential Savings: Moderate (2% to 5%)

Description: Blowdown water temperature is usually over 300° F depending on the pressure and it contains a lot of energy that can be wasted if it is not put back to work in the boiler somehow. About 15% of the blowdown water will flash to low pressure steam so it is a very good source of low pressure steam and is usually used in the deareator/feed water heater. This steam can be recovered in a flash tank and the rest of the heat in a heat exchanger. If steam is not needed, than a simpler heat exchanger recovery configuration can be used to heat feed water.

Advantages/Disadvantages: Proven technology.

38. Automatic Blowdown Control

Cost: Moderate

Potential Savings: Moderate (2%-3%)

Description: When high temperature, high energy blowdown water is not closely controlled, energy is wasted by dumping too much water out of the boiler at one time (**Fig 10.21**). This water then must be replaced by cold make up water. Lack of control may also cause the dissolved solids concentration to rise above the scaling level. Automatic control eliminates these problems and saves water, chemicals and energy.

Advantages/Disadvantages: This option requires good reliable equipment which must be kept calibrated.

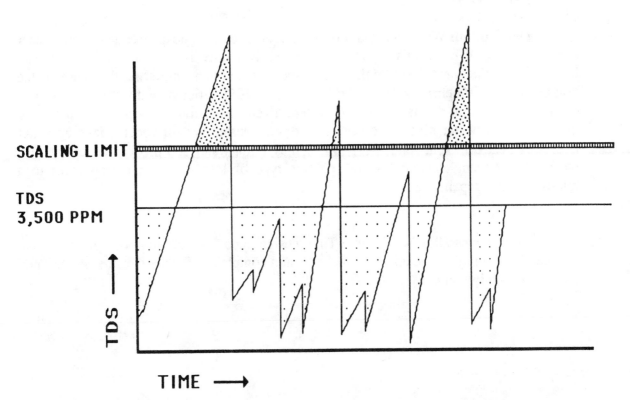

Figure 10.21, Automatic blowdown control eliminates high concentrations of impurities and waste of heat by excessive dumping of hot boiler water.

Over 50 Ways to Improve Boiler Efficiency

39. Use Blowdown to Keep Idle Boilers Warm

Cost: Low

Savings: Low

Description: If you are using steam to protect idle boilers from freezing or need to keep them warm, hot blowdown water can be diverted through idle boilers. This is a tricky option and boiler water levels and chemical build up must be watched carefully. An alternative is to use a heat exchanger with a circulating system.

Advantages/Disadvantages: This is a good use of low grade energy. First cost, maintenance and operating costs are involved.

40. Vent Condensers

Cost: Moderate

Potential Savings: Moderate

Description: Wherever steam is being vented to the atmosphere, a vent condenser can be used to capture this high level energy. Not only is the energy in the steam involved, but the energy lost by the boiler and steam system in getting the steam to the point where it is being lost. The boiler may be 70% efficient with a 30% loss just producing the steam that is being lost through a vent. Locations for vent condensers are deareators, feedwater heaters, receivers and condensate system vents. Make up water is usually heated in vent condensers, but domestic hot water and process water can also be heated. A vent condenser can keep this type of loss from large trap leaks to a minimum until repairs can be made.

Advantages/Disadvantages: The design and installation of piping, heat exchangers and pumps are an additional cost and complexity for the steam system. They are proven energy savers.

Over 50 Ways to Improve Boiler Efficiency

41. Use Heat Pumps with Source Heat From Waste Energy from Boiler Plant

Cost: Moderate

Savings: Moderate

Description: There is a lot of low grade waste heat available from boilers and distribution systems. Two examples are the flue gas stream and the blowdown system. This heat can be put to work at a higher temperature by the use of heat pumps. One system design circulated 150 ° F water from a waste heat recovery system throughout a plant using heat pumps to raise the water temperature for local heating and process demands.

The key to how this operates is in the term: Coefficient of Performance (COP), a measure of the ratio of energy output to energy input. Some heat pump units have a COP of over 5 which means the energy to operate them is only about 20% of the energy gained.

Advantages/Disadvantages: First cost, operation and maintenance expenses involved. It is a good way to recover low grade waste heat.

42. Increasing Cycles Of Concentration of Boiler Water

Cost: Low

Potential Savings: Moderate

Description: Cycles of concentration is the ratio of impurities being maintained in the boiler water divided by the impurity level of the boiler feed water (**Figure 10.22**). This level of concentration is regulated by the percentage of blowdown. If a low level of concentration is maintained, then the blowdown rate could be very high, wasting energy. High cycles of concentration require less blowdown and less energy for make up water heating.

Advantages/Disadvantages: The higher the cycles of concentration to be maintained, the closer the scale formation limit is approached so good control is required.

If this process is not closely regulated, there is a chance for scale to form in the boiler which could lead to tube failure or costly cleaning. The general practice has been to stay away from higher levels of concentration if good control could not be assured, this can waste energy so the pros and cons should be carefully evaluated. Feed water purity is

163

another issue which must be evaluated to really understand the situation. If make-up feed water has high impurities, then higher cycles of concentration will be approached more quickly whereas very pure feed water allows very high cycles of concentration, eliminating blowdown losses and make up water heating to a large degree.

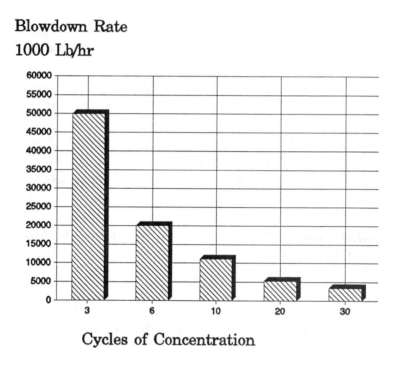

Blowdown Rate 1000 Lb/hr

Cycles of Concentration

Figure 10.22, Increasing Cycles of Concentration. The purer the feed water, the less blowdown required and the higher cycles of concentration. For example, if feedwater quality is improved to increase the concentration from 3 to 6 times, the blowdown rate for a 100,000 lb/hr boiler is reduced from 33.3 to 16.7 percent, or 29,877 lb/hr.

43. Condensing Furnaces, Water Heaters & Boilers

Cost: High

Potential Savings: High

Description: A greater percentage of the waste energy in the exit flue gas stream is in the form of superheated steam formed from the combustion of hydrogen in the fuel and other moisture. For every pound of hydrogen that is burned with the fuel, about nine pounds of water is formed which escapes out of the stack as superheated vapor carrying with it a great deal of energy. It is difficult to recover the latent heat from this vapor without dropping the flue gas temperature below the acid dew point temperature **(Figure 10.23)**.

The acid dew point temperature is the temperature at which sulfuric acid and other acids form. An efficiency increase of 11% to 15% is possible with most boilers if the flue gasses can be cooled below the dew point. Natural gas, because of its higher hydrogen content, produces higher efficiency increases with this method than oil.

When flue gas temperatures can be dropped below the dew point, thereby extracting latent heat from the flue gas moisture, efficiencies above 95% can be expected.

Examples of this technology are: pulse combustion, advanced fiber burner designs, special coated plate exchangers, glass tube exchangers, teflon coated tubes and spray tower direct-contact heat exchangers.

Advantages/Disadvantages: This emerging technology is the only way to get efficiencies in the mid 90% range. Special designs are required, most installations have been highly successful. Acid formation does not seem to be a problem in proper designs.

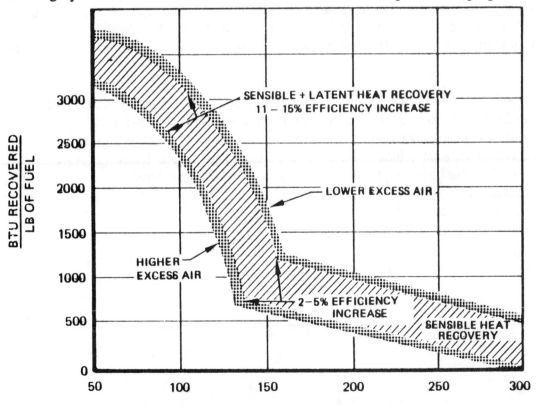

Figure 10.23, Btu recovery from condensing flue gases.

Over 50 Ways to Improve Boiler Efficiency

Figure 10.23 demonstrates the heat recovery potential from a natural gas fired boiler. Reducing exit gas temperature to below the condensing point where latent heat is given up by the flue gas increases heat recovery dramatically. The potential to recover heat depends on the hydrogen ratio of the fuel and excess air. Typically for natural gas boilers, an 11 to 15 percent efficiency increase is possible.

44. Direct Contact Condensation Heat Recovery

Cost: High

Potential Savings: High - 8% to 20%

Description: The direct contact heat recovery system is very efficient because the water used for heat recovery makes direct contact with the flue gasses, eliminating thermal transfer barriers found in indirect heat exchange processes. In the direct contact condensation heat recovery tower, water is sprayed in the top and flue gas comes in from near the bottom. Three common tower types are in general use **(Figure 10.24)**, the packed tower, the baffle-tray tower and the open spray tower.

The hot water from the tower is circulated through a heat exchanger and then back to the spray tower **(Figure 10.25)**. Acid has not been a problem with these units because of dilution from the large quantity of water in the condensed flue gases. The water at approximately 110°F to 130°F can be used for process applications and domestic hot water.

Advantages/Disadvantages: The temperature of the water available from this process is 110° F to 130° F. Efficiencies approaching 97% are possible. Additional forced draft fan capacity may be needed to compensate for the loss of chimney effect due to the cooler exhaust temperatures and the resistance of the recovery system. Acid formation has not been a significant problem with natural gas. Units for oil fired boilers have been developed and are feasible. Very few units have been designed and tested for coal and heavy residual oil.

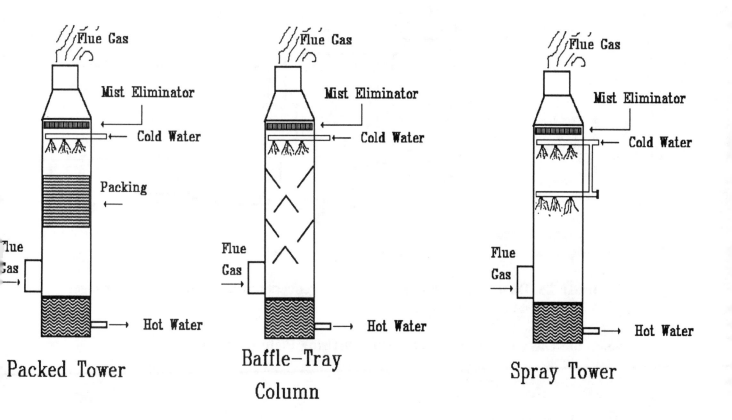

Figure 10.24, direct contact heat recuperation capturing latent heat from flue gases in the condensing range. Packed tower type has high delta-p (5" to 10" W.C.) with high water temperature of 150 F. Baffle-tray column type has medium delta-p (< 1" W.C.) with water temperature of approximately 120 F. The spray type tower has low delta-p (< 1" W.C.) with a water temperature of approximately 120 F.

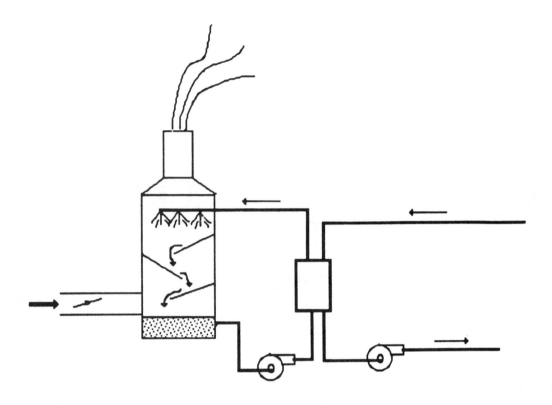

Figure 10.25, Direct contact condensation heat recovery tower unit with heat exchanger.

45. Flue Gas Condensing Heat Recovery Using Plate Exchangers

Cost: Moderate

Potential Savings: Moderate to High

Description: Plate type heat exchangers can be used to recover waste heat for air preheating as well as feed water heating. Because the plates can be fabricated from materials such as stainless steel, titanium and teflon coated steel, units can be designed to bring exhaust gas temperatures well below the acid dew point. This allows the recovery of latent heat from the flue gas, increasing efficiency from 5 to 15 percent. One manufacturer has made a breakthrough with their plate exchanger which is not welded, but torqued together with resilient sections. This type of construction allows the unit to experience thermal expansion without cracking welds or causing other stress related damage.

Advantages/Disadvantages: Materials must withstand acid condensate. Efficiencies approaching 97% possible.

46. Waste Heat for Process Applications

Cost: Moderate

Potential Savings: High

Description: If waste heat can not be recycled into the boiler process, then it can be transferred for other applications like water heating and warm air for drying processes. The best choice is to return the heat to the boiler itself, to eliminate load matching and other complications.

Advantages/Disadvantages: The complexity of transporting the heat to its point of use may be expensive. When waste heat is returned to the boiler, automatic load matching occurs. When using waste heat for process loads, the demand may not always match the boiler loads and special controls may be needed.

47. Submerged Combustion

Cost: Moderate

Potential Savings: High

Description: The combustion process takes place under-water with direct contact heat exchange. Typical use is for heating swimming pools and laundry water (**Figure 10.26**).

Advantages/Disadvantages: Fuel and air must be pressurized to overcome submersion. Efficiencies approaching 95% are possible.

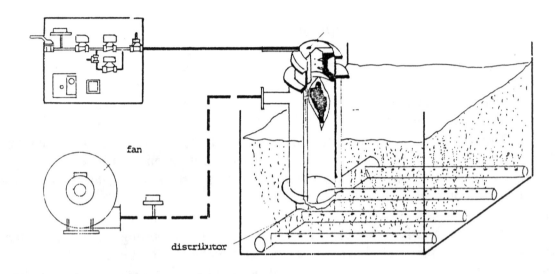

Figure 10.26. Submerged combustion water heating unit capable of efficiencies in the mid 90% range.

48. Solar Augmentation for Water Heating

Cost: Moderate

Savings: Moderate

Description: The **rule of thumb** of each 10 degrees that you can heat the water going into a boiler will improve efficiency by 1%, applies in this case. In some rare applications, make up feed water is a good candidate for solar augmentation. Also, domestic hot water heating efficiency can be improved by this option. Many studies have shown that this option is seldom cost effective, but there are applications where it can pay off.

Advantages/Disadvantages: This application would be most unusual, but possible. Financial analysis would be required for justification.

Over 50 Ways to Improve Boiler Efficiency

49. Satellite Boilers

Cost: Moderate to High

Potential Savings: Very High

Description: Distribution system losses can be very high and the best way to eliminate distribution system losses is to eliminate the distribution system. Originally, central plants were designed around the use of coal because it was the only fuel available and now many facilities have miles of old and questionable distribution systems which were built before the day of the reliable and safe satellite boiler.

When you see brown grass above buried steam lines in the summer and melted snow over the distribution system in the winter, you are probably looking at heat losses from buried steam and condensate piping.

Every foot of piping is a possible source of leaks, energy losses, acid corrosion and other maintenance problems. Rain soaked and degraded insulation, flooded systems, steam plumes along the steam mains in streets and fields and cold or lost condensate are all indicators of major distribution system problems.

The central steam plant can be a huge liability, one only has to analyze steam production during warm weather or holidays to discover that large losses exist. Reliable and safe control technology and the availability oil and gas supplies now make individual satellite boilers a safe and efficient way to avoid large distribution system losses.

Advantages/Disadvantages: The cost of installing satellite boilers can be quite high, but the savings can be very high in both maintenance costs and in energy savings.

50. High Temperature Hot Water Systems

Cost: High

Potential Savings: High

Description:

(1) High temperature hot water has a system temperature from 350° F to 450° F. Medium temperature hot water is between 250° F and 350° F and Low temperature hot water is below 250° F.

(2) One cubic foot of water at 350° F contains 50 times as much heat as a cubic foot of saturated steam at 350° F.

(3) When a high temperature hot water system cools a cubic foot of water from 350° F to 250° F, it releases approximately 27.6 times more heat than a cubic foot of saturated steam at 350° F.

Advantages/Disadvantages: Losses with a high temperature hot water system are less than steam because:

(a) It is a closed system and leaks are usually less than one half of one percent whereas a steam system can require up to 100% make up feed water. The actual requirement for makeup water can vary widely.

(b) It minimizes losses associated with steam traps, valve stem and packing gland leakage, blowdown and flash losses associated with open condensate receivers.

(c) It can pack more usable heat into a given volume than a steam system and needs smaller diameter piping.

(d) Return water temperature must be maintained above a certain limit to avoid thermal stress and piping system damage.

A recent study has shown that in a 20 Million Btu/hr system, comparing HTW at 350° F to saturated steam, showed that there is a $300,000 a year savings with HTW. Over a life cycle of 25 years the savings was over $14 million.

51. Recover Hot Condensate

Cost: Could be high but is basically a maintenance item.

Potential Savings: High

Description: The loss of hot condensate wastes energy and money in several ways:

(a) The condensate should return at 160° F - 190° F. If it doesn't return, this heat is lost.

(b) Makeup feed water must be chemically treated and heated from 50° F to 160° F - 190° F to replace the lost condensate.

Over 50 Ways to Improve Boiler Efficiency

(c) Condensate is essentially pure water. Having to replace it with "makeup" water containing impurities increases the amount of blowdown required.

(d) Chemical treatment of the makeup feed water and the cost of the water itself can be avoided.

There is an additional factor here, and that is if the condensate system has leaks or is not functional and steam traps have failed and are blowing steam, all of this high energy steam is also disappearing.

Electric pumps have an uncertain lifetime in environments which occasionally can flood, also centrifugal pumps have difficulty handling condensate near the boiling point. These electrical units can be replaced with pump sets which use steam pressure to pump condensate back to the boiler plant.

Advantages/Disadvantages: Repair and replacement of condensate return systems can be quite expensive. The economic gain or savings available from keeping condensate systems well maintained is also very high.

52. Steam Trap Maintenance

Cost: Low

Potential Savings: High

Description: Steam traps can waste a tremendous amount of energy; even when working properly steam traps are inefficient when you consider the flash losses. Traps go through millions of cycles a year and develop leaks and many types have high failure rates. Many studies have shown that large savings are possible with proper steam trap maintenance programs.

Many condensate systems are "open" which means that they vent to atmosphere. Because of the high pressure on the working side of the trap and the low pressure on the outlet side, 10% to 16% of the condensate going through healthy traps will flash into steam. Add to this the many problems found in traps, like closing slowly, allowing steam to escape, leaky seats or just not closing at all and it becomes obvious there is a potential for large hidden steam losses, maybe over 50%. Extensive studies have shown that in most plants, a very high percentage of traps have failed one way or the other. The real question is, "How much of this expensive steam is really being put to productive use and how much is being lost?" When the potential for loss is documented, it will become be self evident that a good maintenance program is required.

Over 50 Ways to Improve Boiler Efficiency

Advantages/Disadvantages: Until the potential savings of good steam trap maintenance is realized by management, big energy dollars will be silently lost in distribution systems.

53. Improve Insulation

Cost: Moderate

Potential Savings: Payback one to two years.

Description: Boilers and steam systems have surface temperatures from 350° F to 450° F and above and must have effective insulation to prevent heat from escaping from boiler and piping surfaces **(Figure 10.27)**. Often insulation is removed from valves, piping and boiler parts , for repairs and not replaced. This bare surface is a potential safety hazard as well as a source of lost energy. One danger associated with poor insulation is that fuel and oil leaks have been known to ignite off these hot surfaces and start fires.

Advantages/Disadvantages: Asbestos may be encountered in older systems. Insulating bare steam piping is usually a quick pay back item.

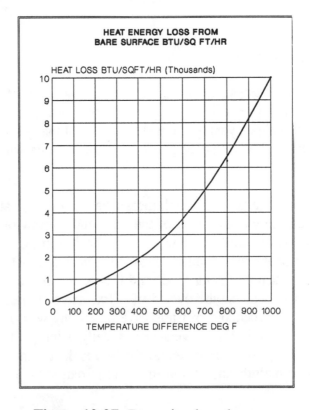

Figure 10.27, Bare pipe heat losses.

Over 50 Ways to Improve Boiler Efficiency

54. Modulate Vacuum Pressure in Vacuum Return Systems

Cost: Original Design (low)

Savings: High

Description: There are some older steam systems around that are of the condensing type, where steam is circulated back to a condensing unit in the boiler room. By modulating the steam pressure, the temperature of the distribution system can be matched to the demand for a particular weather condition. On a cold day the steam pressure and temperature can be allowed to rise and on warmer days the pressure and steam temperature can be controlled below atmospheric and temperatures below 212°F.

55. Reduce Steam Pressure

Cost: Low

Potential Savings: Moderate depending on the condition of the system.

Description: Lower the steam system pressure set point.

Higher steam operating pressures result in higher losses because of higher stack temperatures, larger steam leaks, higher boiler and distribution insulation losses and higher steam trap losses.

Advantages/Disadvantages: Savings may be available but this option must be well thought out and engineered because of the many hidden problems that it can create.

Lowering steam pressure reduces these losses but can create some serious problems in the plant:

Disadvantages

(a) Steam volume increase may cause excessive piping velocities. Volume of steam increases rapidly as pressure is reduced.

4.4 Ft³/lb	100 PSI
8.6 Ft³/lb	50 PSI
25.0 Ft³/lb	15 PSI

Over 50 Ways to Improve Boiler Efficiency

(b) Temperature sensitive traps may not work properly with lowered steam temperature.

(c) Pressure change may change the calibration of installed instruments, increased turbulence in pipe from higher velocities may also affect instruments and meters.

(d) Steam driven equipment like emergency generators may not have required pressure to operate. If they can operate, efficiency and capacity will be lower.

(e) Cavation noises may develop in feed water valves.

(f) Atomizing steam pressure may be too low.

(g) Lowered steam temperature may be too low for the designed range of heat exchangers.

(h) Valves may become noisy and whistle because of higher velocities through the restricted areas of valve internals.

(i) Differential pressure change across steam traps may change performance.

(j) Higher volume of steam leaving boiler may cause priming (water entering piping from boiler).

(k) Higher volume of water may affect reserve pumping capacity of original design.

Advantages

(a) Lower boiler temperatures will result in lower stack temperatures and lower boiler skin losses.

(b) Lower heat losses in distribution system piping.

(c) Steam leaks will be less.

(d) Flash steam losses will be reduced.

Over 50 Ways to Improve Boiler Efficiency

56. Steam Compression

Cost: Moderate to High

Potential Savings: High

Description: If you have low pressure waste steam or if you need a small amount of higher pressure steam at some remote location in the distribution system, then steam compression may serve to your advantage. With a source of low pressure steam that would otherwise be wasted, it can be salvaged and brought back up to working pressure without large losses by using steam compression. In cases where a large distribution system must be maintained at a high pressure to just serve a small high pressure load, the pressure for the system can be reduced and the small high pressure demand can be met with steam compression. Screw compressors and steam eductor sets have been developed for steam compression which are driven by either electric or engine type prime movers.

Advantages/Disadvantages: Using steam compression to satisfy requirements in otherwise low pressure systems makes sense.

57. Install Pressure Reducing Valves

Cost: Moderate

Potential Savings: Moderate

Description: Installing pressure reducing valves near the point-of-use can reduce flash steam losses. Also, high pressure steam has a higher temperature, lower specific volume and is economical to distribute through steam mains requiring smaller sized piping.

As a general rule, steam is most economically distributed at high pressure and if reduced to the lowest pressure that will satisfy the temperature requirements at the point-of-use, the least amount of energy will be lost.

Advantages/Disadvantages: The cost of installing and maintaining pressure reducing valves and possible replacement of steam traps will have to be balanced against predicted savings.

58. Flash Steam Utilization

Cost: Moderate

Potential Savings: Moderate

Description: Use a flash steam recovery vessel to collect condensate from high pressure sources. It separates steam from condensate and the low pressure flash steam is piped off to some useful purpose.

Only as a last resort should flash steam be vented to the atmosphere and lost.

Advantages/Disadvantages: New coils or heat exchangers may be required to utilize the flash steam for heating and other uses. Flash steam possibilities are becoming recognized.

59. Temperature Set Back

Cost: Low

Potential Savings: Moderate

Description: Temperature setback uses outside air temperature to shut the boiler off or reset the circulating water temperature.

Advantages/Disadvantages: Additional costs involved, temperature fluctuations in distribution system should be set so as not to damage the piping system with abrupt temperature changes.

Over 50 Ways to Improve Boiler Efficiency

60. Adjustable Speed Drives

Cost: Moderate

Potential Savings: Moderate to high.

Description: When pumps and fans operate at higher rpms than required for a particular load, electrical energy is lost. The pump and fan laws apply which states that the energy used varies with the cube of the rpm. Large savings can be accomplished by reducing the rpm of rotating equipment. Variable speed drives are a new dimension in the cost effective management of plants. For example, if the speed of a pump or fan is reduced by one half, there is a 88% reduction in energy use.

Advantages/Disadvantages: Several plants have successfully applied adjustable speed drive to boiler operations. Adjustable Speed Drive (ASD) is becoming a proven way to save electrical costs.

61. Use Steam to Drive Equipment Such as Pumps and Blowers

Cost: Moderate to high.

Potential Savings: Moderate to high.

Description: Electrically driven pumps and blowers are usually installed because they are simpler and require less maintenance, however when the cost of electricity is considered, steam driven auxiliaries may be more economical. If there is a place to utilize the full volume of exhaust steam from steam driven equipment, then a situation similar to cogeneration exists, where the rejected heat can be utilized improving over all cycle efficiency. Steam driven equipment is especially attractive where electrical demand charges are very high.

Advantages/Disadvantages: High first cost, maintenance requirements and the need for more operator skill work against this option. Energy and cost savings could be very significant.

Over 50 Ways to Improve Boiler Efficiency

62. Use Soft Firebrick in Combustion Chambers

Cost: Moderate (usually a design choice)

Potential Savings: Moderate

Description: If there is a choice in combustion chamber materials for original purchase, for redesign or for repairs, light weight fire brick or ceramic fiber material should be considered. This material heats up faster taking the combustion zone through its cold smoky period more quickly **(Figure 10.28)**. Its use produces inherently lower smoke spot numbers on light off. This should be considered for low use boilers.

Advantages/disadvantages: The obvious disadvantage is high first cost if retrofit is required. The use of light refractory improves combustion performance on initial light off and soaks up less heat on/off cycles.

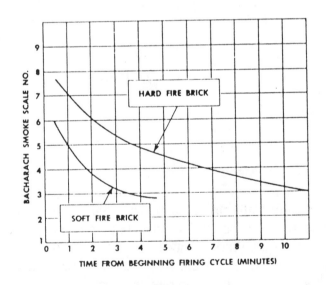

Figure 10.28, The effect of combustion chamber material on smoke.

180

Over 50 Ways to Improve Boiler Efficiency

63. Convert to Infra-Red Heaters

Cost: Moderate

Potential Savings: Moderate

Description: Infra-Red heaters have a high combustion efficiency and an ability to heat selected line of sight areas efficiently, such as hangars, warehouses, gymnasiums and maintenance facilities. They put the heat where it's needed and do not heat up a large volume of air that may uselessly float up to the top of a large open building.

Advantages/disadvantages: The Infra-Red heater is usually self contained and mounts in the overhead, out of the way. They may be an intelligent replacement to old steam, hot water and hot air systems.

64. Use Flue Gas as a source of Carbon Dioxide

Cost: Moderate

Potential Savings: Moderate

Description: In some locations flue gas, with its high CO_2 content, has been used as a source of CO_2 for various purposes. This is a novel way to use a "waste stream".

Advantages/disadvantages: There would need to be a market for the CO_2 and the boiler would have to be a reliable source.

65. Intermittent Ignition Devices

Cost: Low

Savings: Moderate

Description: It doesn't make any sense to keep a flame lit in a furnace or boiler if heat isn't needed; it wastes energy. Especially with gas fired equipment, there are many smaller sized boilers and furnaces with continuous pilot flames. On a 24 hour a day basis, year after year, these small flames can use a lot of Btus. Intermittent ignition technology has been proven to be reliable and safe, so these older continuous pilot lights should be replaced with Intermittent Ignition Devices.

Advantages/Disadvantages: Intermittent ignition devices are proven, reliable equipment. They have been designed to retrofit on most boilers and furnaces and are not expensive. Their payback period is relatively short.

Chapter 11

Boiler Plant Calculations

Finding excess air from flue gas readings.

The purpose of this section is to show how excess air levels are determined by measuring oxygen (O_2) and carbon dioxide (CO_2), and using the Combustion Efficiency Tables. (pp. 186 - 189)

1. The measured oxygen level in a natural gas boiler is 8%, what is the excess air?

    Answer:

2. The measured CO_2 level for a boiler burning natural gas is 9%, what is the excess air?

    Answer:

3. The measured CO_2 level for a boiler burning no. 6 fuel oil is 15%, what is the excess air?

    Answer:

4. Rule of thumb. Using the oxygen scale, can you develop a rule of thumb relationship for excess air for oxygen levels between 0% and 4%.

    Answer:

5. Why is the maximum CO_2 level at 0% excess air higher for no. 6 fuel oil than for Natural
Gas?

    Answer:

[Answers: **1.** 54%, **2.** 27%, **3.** 10%, **4.** 5 X O2 level = Excess air, **5.** There is more carbon in heavier fuels.]

EFFICIENCY
(Heat Loss Method)

The purpose of this section is to learn how to determine combustion efficiency using prepared tables.

Determine Heat Loss Efficiency of the following examples using the combustion tables. (pp. 186 - 189)

Example 1.

Fuel Type	No. 6
Carbon Dioxide	9.1%
Stack Temperature	560F
Air Temperature	60F
Net Temperature	500F

Efficiency _____

Example 2.

Fuel Type	No. 2
Oxygen	9%
Stack Temperature	420F
Air Temperature	80F
Net Temperature	340F

Efficiency _____

Example 3.

Fuel Type	Natural Gas
Carbon Dioxide	8.9%
Stack Temperature	580F
Air Temperature	60F
Net Temperature	520F

Efficiency _____

Example 4.

Fuel Type	Wood 30% Moisture
Oxygen	10%
Stack Temperature	680F
Air Temperature	80F
Net Temperature	600F

Efficiency _____

[Answers **1.** 79.3%, **2.** 83.5%, **3.** 79.4%, **4.** 70%]

Sample

Heat Loss Tables

for

Natural Gas

No. 2 Fuel Oil

No. 6 Fuel Oil

and

Wood 30% Moisture

Boiler Plant Calculations

NATURAL GAS

EXIT GAS HEAT LOSSES

% EXCESS AIR	% OXYGEN	% CO2	450	460	470	480	490	500	510	520	530	540	550	560	570	580	590	600
0.0	0.0	11.7	17.0	17.1	17.3	17.4	17.6	17.8	17.9	18.1	18.3	18.4	18.6	18.7	18.9	19.1	19.2	19.4
2.2	0.5	11.4	17.1	17.3	17.5	17.6	17.8	18.0	18.1	18.3	18.5	18.6	18.8	19.0	19.1	19.3	19.5	19.6
4.4	1.0	11.1	17.3	17.5	17.6	17.8	18.0	18.2	18.3	18.5	18.7	18.8	19.0	19.2	19.3	19.5	19.7	19.9
6.8	1.5	10.9	17.5	17.7	17.8	18.0	18.2	18.4	18.5	18.7	18.9	19.1	19.2	19.4	19.6	19.8	19.9	20.1
9.3	2.0	10.6	17.7	17.9	18.1	18.2	18.4	18.6	18.8	18.9	19.1	19.3	19.5	19.7	19.8	20.0	20.2	20.4
12.0	2.5	10.3	17.9	18.1	18.3	18.5	18.6	18.8	19.0	19.2	19.4	19.6	19.7	19.9	20.1	20.3	20.5	20.7
14.8	3.0	10.0	18.1	18.3	18.5	18.7	18.9	19.1	19.3	19.4	19.6	19.8	20.0	20.2	20.4	20.6	20.8	21.0
17.7	3.5	9.8	18.4	18.6	18.8	18.9	19.1	19.3	19.5	19.7	19.9	20.1	20.3	20.5	20.7	20.9	21.1	21.3
20.8	4.0	9.5	18.6	18.8	19.0	19.2	19.4	19.6	19.8	20.0	20.2	20.4	20.6	20.8	21.0	21.2	21.4	21.6
24.1	4.5	9.2	18.9	19.1	19.3	19.5	19.7	19.9	20.1	20.3	20.5	20.7	20.9	21.1	21.3	21.5	21.7	22.0
27.6	5.0	8.9	19.2	19.4	19.6	19.8	20.0	20.2	20.4	20.6	20.8	21.1	21.3	21.5	21.7	21.9	22.1	22.3
31.4	5.5	8.6	19.5	19.7	19.9	20.1	20.3	20.5	20.8	21.0	21.2	21.4	21.6	21.9	22.1	22.3	22.5	22.7
35.4	6.0	8.4	19.8	20.0	20.2	20.4	20.7	20.9	21.1	21.3	21.6	21.8	22.0	22.2	22.5	22.7	22.9	23.1
39.6	6.5	8.1	20.1	20.3	20.6	20.8	21.0	21.3	21.5	21.7	22.0	22.2	22.4	22.7	22.9	23.1	23.4	23.6
44.2	7.0	7.8	20.5	20.7	21.0	21.2	21.4	21.7	21.9	22.2	22.4	22.6	22.9	23.1	23.4	23.6	23.8	24.1
49.0	7.5	7.5	20.9	21.1	21.4	21.6	21.9	22.1	22.4	22.6	22.9	23.1	23.4	23.6	23.9	24.1	24.4	24.6
54.3	8.0	7.2	21.3	21.6	21.8	22.1	22.3	22.6	22.8	23.1	23.4	23.6	23.9	24.1	24.4	24.7	24.9	25.2
60.0	8.5	7.0	21.7	22.0	22.3	22.6	22.8	23.1	23.4	23.6	23.9	24.2	24.4	24.7	25.0	25.2	25.5	25.8
66.1	9.0	6.7	22.2	22.5	22.8	23.1	23.4	23.6	23.9	24.2	24.5	24.8	25.0	25.3	25.6	25.9	26.2	26.4
72.8	9.5	6.4	22.8	23.1	23.4	23.7	23.9	24.2	24.5	24.8	25.1	25.4	25.7	26.0	26.3	26.6	26.9	27.2
80.0	10.0	6.1	23.4	23.7	24.0	24.3	24.6	24.9	25.2	25.5	25.8	26.1	26.4	26.7	27.0	27.3	27.6	27.9
88.0	10.5	5.9	24.0	24.3	24.6	25.0	25.3	25.6	25.9	26.2	26.6	26.9	27.2	27.5	27.8	28.2	28.5	28.8
96.7	11.0	5.6	24.7	25.0	25.4	25.7	26.0	26.4	26.7	27.1	27.4	27.7	28.1	28.4	28.7	29.1	29.4	29.7
106.3	11.5	5.3	25.5	25.8	26.2	26.5	26.9	27.3	27.6	28.0	28.3	28.7	29.0	29.4	29.7	30.1	30.4	30.8
117.0	12.0	5.0	26.4	26.7	27.1	27.5	27.8	28.2	28.6	29.0	29.3	29.7	30.1	30.4	30.8	31.2	31.6	31.9
128.9	12.5	4.7	27.3	27.7	28.1	28.5	28.9	29.3	29.7	30.1	30.5	30.9	31.3	31.6	32.0	32.4	32.8	33.2
142.3	13.0	4.5	28.4	28.8	29.2	29.7	30.1	30.5	30.9	31.3	31.7	32.2	32.6	33.0	33.4	33.8	34.3	34.7
157.4	13.5	4.2	29.6	30.1	30.5	31.0	31.4	31.9	32.3	32.8	33.2	33.6	34.1	34.5	35.0	35.4	35.9	36.3
174.6	14.0	3.9	31.1	31.5	32.0	32.5	33.0	33.4	33.9	34.4	34.9	35.3	35.8	36.3	36.8	37.2	37.7	38.2
194.4	14.5	3.6	32.7	33.2	33.7	34.2	34.7	35.2	35.8	36.3	36.8	37.3	37.8	38.3	38.8	39.3	39.9	40.4
217.4	15.0	3.3	34.6	35.1	35.7	36.2	36.8	37.4	37.9	38.5	39.0	39.6	40.1	40.7	41.2	41.8	42.3	42.9
244.4	15.5	3.1	36.8	37.4	38.0	38.6	39.2	39.8	40.5	41.1	41.7	42.3	42.9	43.5	44.1	44.7	45.3	45.9
276.7	16.0	2.8	39.5	40.2	40.8	41.5	42.2	42.8	43.5	44.2	44.8	45.5	46.2	46.8	47.5	48.2	48.8	49.5

NET STACK TEMPERATURE DEG F
EXIT FLUE GAS TEMPERATURE - COMBUSTION AIR TEMPERATURE

Higher Heating Value (HHV)   21,830   Btu/lb

Carbon	69.4%
Hydrogen	22.5%
Ultimate CO2	11.7%

(450 To 600 F)
NATURAL GAS

186

Boiler Plant Calculations

No. 2 Fuel Oil

EXIT GAS HEAT LOSSES

% EXCESS AIR	% OXYGEN	% CO2	NET STACK TEMPERATURE DEG F EXIT FLUE GAS TEMPERATURE - COMBUSTION AIR TEMPERATURE															
			300	310	320	330	340	350	360	370	380	390	400	410	420	430	440	450
0.0	0.0	15.7	11.5	11.7	11.9	12.1	12.2	12.4	12.6	12.8	12.9	13.1	13.3	13.5	13.7	13.8	14.0	14.2
2.3	0.5	15.3	11.6	11.8	12.0	12.2	12.4	12.6	12.7	12.9	13.1	13.3	13.5	13.7	13.8	14.0	14.2	14.4
4.6	1.0	15.0	11.8	12.0	12.1	12.3	12.5	12.7	12.9	13.1	13.3	13.5	13.6	13.8	14.0	14.2	14.4	14.6
7.1	1.5	14.6	11.9	12.1	12.3	12.5	12.7	12.9	13.1	13.3	13.4	13.6	13.8	14.0	14.2	14.4	14.6	14.8
9.8	2.0	14.2	12.1	12.2	12.4	12.6	12.8	13.0	13.2	13.4	13.6	13.8	14.0	14.2	14.4	14.6	14.8	15.0
12.5	2.5	13.8	12.2	12.4	12.6	12.8	13.0	13.2	13.4	13.6	13.8	14.0	14.2	14.4	14.6	14.8	15.0	15.2
15.5	3.0	13.5	12.4	12.6	12.8	13.0	13.2	13.4	13.6	13.8	14.0	14.2	14.4	14.6	14.8	15.1	15.3	15.5
18.5	3.5	13.1	12.5	12.7	13.0	13.2	13.4	13.6	13.8	14.0	14.2	14.4	14.7	14.9	15.1	15.3	15.5	15.7
21.8	4.0	12.7	12.7	12.9	13.1	13.4	13.6	13.8	14.0	14.2	14.5	14.7	14.9	15.1	15.3	15.6	15.8	16.0
25.3	4.5	12.3	12.9	13.1	13.4	13.6	13.8	14.0	14.2	14.5	14.7	14.9	15.1	15.4	15.6	15.8	16.0	16.3
29.0	5.0	12.0	13.1	13.3	13.6	13.8	14.0	14.3	14.5	14.7	15.0	15.2	15.4	15.6	15.9	16.1	16.3	16.6
32.9	5.5	11.6	13.3	13.6	13.8	14.0	14.3	14.5	14.7	15.0	15.2	15.5	15.7	15.9	16.2	16.4	16.7	16.9
37.0	6.0	11.2	13.5	13.8	14.0	14.3	14.5	14.8	15.0	15.3	15.5	15.8	16.0	16.3	16.5	16.7	17.0	17.2
41.5	6.5	10.8	13.8	14.0	14.3	14.6	14.8	15.1	15.3	15.6	15.8	16.1	16.3	16.6	16.8	17.1	17.4	17.6
46.2	7.0	10.5	14.1	14.3	14.6	14.8	15.1	15.4	15.6	15.9	16.2	16.4	16.7	16.9	17.2	17.5	17.7	18.0
51.4	7.5	10.1	14.3	14.6	14.9	15.2	15.4	15.7	16.0	16.2	16.5	16.8	17.1	17.3	17.6	17.9	18.1	18.4
56.9	8.0	9.7	14.6	14.9	15.2	15.5	15.8	16.1	16.3	16.6	16.9	17.2	17.5	17.7	18.0	18.3	18.6	18.9
62.8	8.5	9.3	15.0	15.3	15.6	15.8	16.1	16.4	16.7	17.0	17.3	17.6	17.9	18.2	18.5	18.8	19.1	19.4
69.2	9.0	9.0	15.3	15.6	15.9	16.2	16.5	16.8	17.2	17.5	17.8	18.1	18.4	18.7	19.0	19.3	19.6	19.9
76.2	9.5	8.6	15.7	16.0	16.3	16.7	17.0	17.3	17.6	17.9	18.3	18.6	18.9	19.2	19.5	19.8	20.2	20.5
83.8	10.0	8.2	16.1	16.5	16.8	17.1	17.5	17.8	18.1	18.5	18.8	19.1	19.4	19.8	20.1	20.4	20.8	21.1
92.1	10.5	7.9	16.6	16.9	17.3	17.6	18.0	18.3	18.7	19.0	19.4	19.7	20.1	20.4	20.8	21.1	21.5	21.8
101.2	11.0	7.5	17.1	17.5	17.8	18.2	18.6	18.9	19.3	19.6	20.0	20.4	20.7	21.1	21.5	21.8	22.2	22.6
111.3	11.5	7.1	17.7	18.0	18.4	18.8	19.2	19.6	20.0	20.3	20.7	21.1	21.5	21.9	22.2	22.6	23.0	23.4
122.5	12.0	6.7	18.3	18.7	19.1	19.5	19.9	20.3	20.7	21.1	21.5	21.9	22.3	22.7	23.1	23.5	23.9	24.3
134.9	12.5	6.4	19.0	19.4	19.8	20.3	20.7	21.1	21.5	22.0	22.4	22.8	23.2	23.7	24.1	24.5	24.9	25.4
148.9	13.0	6.0	19.8	20.2	20.7	21.1	21.6	22.0	22.5	22.9	23.4	23.8	24.3	24.7	25.2	25.6	26.1	26.5
164.7	13.5	5.6	20.6	21.1	21.6	22.1	22.6	23.1	23.5	24.0	24.5	25.0	25.5	25.9	26.4	26.9	27.4	27.9
182.7	14.0	5.2	21.7	22.2	22.7	23.2	23.7	24.2	24.8	25.3	25.8	26.3	26.8	27.3	27.8	28.4	28.9	29.4
203.4	14.5	4.9	22.8	23.4	23.9	24.5	25.0	25.6	26.2	26.7	27.3	27.8	28.4	28.9	29.5	30.0	30.6	31.2
227.5	15.0	4.5	24.2	24.8	25.4	26.0	26.6	27.2	27.8	28.4	29.0	29.6	30.2	30.8	31.4	32.0	32.6	33.2
255.8	15.5	4.1	25.8	26.5	27.1	27.8	28.4	29.1	29.7	30.4	31.0	31.7	32.3	33.0	33.7	34.3	35.0	35.6
289.5	16.0	3.7	27.7	28.5	29.2	29.9	30.6	31.3	32.0	32.8	33.5	34.2	34.9	35.6	36.4	37.1	37.8	38.5

Higher Heating Value (HHV)	137,080	Btu/gal	Carbon	87.3%
	18,993	Btu/lb	Hydrogen	12.5%
	7.213	lb/gal	Ultimate CO2	15.7%
SP GR	0.865	60 F	API	30-34

300 - 450 F
No. 2 FUEL OIL

Boiler Plant Calculations

NO 6 FUEL OIL

EXIT GAS HEAT LOSSES

% EXCESS AIR	% OXYGEN	% CO2	450	460	470	480	490	500	510	520	530	540	550	560	570	580	590	600
			NET STACK TEMPERATURE DEG F																

NET STACK TEMPERATURE DEG F
EXIT FLUE GAS TEMPERATURE - COMBUSTION AIR TEMPERATURE

% EXCESS AIR	% OXYGEN	% CO2	450	460	470	480	490	500	510	520	530	540	550	560	570	580	590	600
0.0	0.0	16.7	12.9	13.1	13.2	13.4	13.6	13.8	13.9	14.1	14.3	14.5	14.7	14.8	15.0	15.2	15.4	15.6
2.3	0.5	16.3	13.1	13.2	13.4	13.6	13.8	14.0	14.2	14.3	14.5	14.7	14.9	15.1	15.3	15.4	15.6	15.8
4.7	1.0	15.9	13.3	13.4	13.6	13.8	14.0	14.2	14.4	14.6	14.8	14.9	15.1	15.3	15.5	15.7	15.9	16.1
7.2	1.5	15.5	13.5	13.6	13.8	14.0	14.2	14.4	14.6	14.8	15.0	15.2	15.4	15.6	15.8	16.0	16.1	16.3
9.9	2.0	15.1	13.7	13.9	14.1	14.3	14.5	14.7	14.9	15.1	15.2	15.4	15.6	15.8	16.0	16.2	16.4	16.6
12.7	2.5	14.7	13.9	14.1	14.3	14.5	14.7	14.9	15.1	15.3	15.5	15.7	15.9	16.1	16.3	16.5	16.7	16.9
15.6	3.0	14.3	14.1	14.3	14.6	14.8	15.0	15.2	15.4	15.6	15.8	16.0	16.2	16.4	16.6	16.8	17.0	17.2
18.8	3.5	13.9	14.4	14.6	14.8	15.0	15.2	15.5	15.7	15.9	16.1	16.3	16.5	16.7	16.9	17.2	17.4	17.6
22.1	4.0	13.5	14.7	14.9	15.1	15.3	15.5	15.8	16.0	16.2	16.4	16.6	16.8	17.1	17.3	17.5	17.7	17.9
25.6	4.5	13.1	14.9	15.2	15.4	15.6	15.8	16.1	16.3	16.5	16.7	17.0	17.2	17.4	17.6	17.9	18.1	18.3
29.3	5.0	12.7	15.2	15.5	15.7	15.9	16.2	16.4	16.6	16.9	17.1	17.3	17.6	17.8	18.0	18.3	18.5	18.7
33.3	5.5	12.3	15.6	15.8	16.0	16.3	16.5	16.8	17.0	17.2	17.5	17.7	18.0	18.2	18.4	18.7	18.9	19.2
37.5	6.0	11.9	15.9	16.2	16.4	16.6	16.9	17.1	17.4	17.6	17.9	18.1	18.4	18.6	18.9	19.1	19.4	19.6
42.0	6.5	11.5	16.3	16.5	16.8	17.0	17.3	17.5	17.8	18.1	18.3	18.6	18.8	19.1	19.3	19.6	19.8	20.1
46.8	7.0	11.1	16.7	16.9	17.2	17.5	17.7	18.0	18.2	18.5	18.8	19.0	19.3	19.6	19.8	20.1	20.4	20.6
52.0	7.5	10.7	17.1	17.4	17.6	17.9	18.2	18.5	18.7	19.0	19.3	19.5	19.8	20.1	20.4	20.6	20.9	21.2
57.5	8.0	10.3	17.5	17.8	18.1	18.4	18.7	19.0	19.2	19.5	19.8	20.1	20.4	20.7	20.9	21.2	21.5	21.8
63.6	8.5	9.9	18.0	18.3	18.6	18.9	19.2	19.5	19.8	20.1	20.4	20.7	21.0	21.3	21.6	21.9	22.1	22.4
70.1	9.0	9.5	18.6	18.9	19.2	19.5	19.8	20.1	20.4	20.7	21.0	21.3	21.6	21.9	22.2	22.5	22.8	23.2
77.1	9.5	9.1	19.1	19.5	19.8	20.1	20.4	20.7	21.1	21.4	21.7	22.0	22.3	22.6	23.0	23.3	23.6	23.9
84.8	10.0	8.7	19.8	20.1	20.4	20.8	21.1	21.4	21.8	22.1	22.4	22.8	23.1	23.4	23.8	24.1	24.4	24.8
93.2	10.5	8.4	20.5	20.8	21.2	21.5	21.9	22.2	22.5	22.9	23.2	23.6	23.9	24.3	24.6	25.0	25.3	25.7
102.4	11.0	8.0	21.2	21.6	22.0	22.3	22.7	23.0	23.4	23.8	24.1	24.5	24.9	25.2	25.6	26.0	26.3	26.7
112.6	11.5	7.6	22.1	22.4	22.8	23.2	23.6	24.0	24.4	24.7	25.1	25.5	25.9	26.3	26.7	27.0	27.4	27.8
123.9	12.0	7.2	23.0	23.4	23.8	24.2	24.6	25.0	25.4	25.8	26.2	26.6	27.0	27.4	27.8	28.2	28.6	29.1
136.5	12.5	6.8	24.0	24.5	24.9	25.3	25.7	26.2	26.6	27.0	27.5	27.9	28.3	28.7	29.2	29.6	30.0	30.4
150.7	13.0	6.4	25.2	25.7	26.1	26.6	27.0	27.5	27.9	28.4	28.8	29.3	29.7	30.2	30.6	31.1	31.6	32.0
166.7	13.5	6.0	26.5	27.0	27.5	28.0	28.5	28.9	29.4	29.9	30.4	30.9	31.4	31.8	32.3	32.8	33.3	33.8
184.9	14.0	5.6	28.1	28.6	29.1	29.6	30.1	30.6	31.2	31.7	32.2	32.7	33.2	33.7	34.2	34.8	35.3	35.8
205.8	14.5	5.2	29.8	30.4	30.9	31.5	32.0	32.6	33.1	33.7	34.2	34.8	35.4	35.9	36.5	37.0	37.6	38.1
230.2	15.0	4.8	31.8	32.4	33.0	33.6	34.2	34.8	35.5	36.1	36.7	37.3	37.9	38.5	39.1	39.7	40.3	40.9
258.8	15.5	4.4	34.3	34.9	35.6	36.2	36.9	37.5	38.2	38.8	39.5	40.2	40.8	41.5	42.1	42.8	43.4	44.1
292.9	16.0	4.0	37.2	37.9	38.6	39.3	40.0	40.8	41.5	42.2	42.9	43.6	44.3	45.1	45.8	46.5	47.2	47.9

Higher Heating Value (HHV)	153,120	Carbon	88.5%
	18126 Btu/lb	Hydrogen	9.3%
	8.45	Ultimate CO2	16.7%
SP GR	.946-1.01 a 60 F	API	12 - 16

450 - 600 F
NO 6 FUEL OIL

Boiler Plant Calculations

EXIT GAS HEAT LOSSES

% EXCESS AIR	% OXYGEN	% CO2	600	610	620	630	640	650	660	670	680	690	700	710	720	730	740	750
			NET STACK TEMPERATURE DEG F																

NET STACK TEMPERATURE DEG F
EXIT FLUE GAS TEMPERATURE - COMBUSTION AIR TEMPERATURE

% EXCESS AIR	% OXYGEN	% CO2	600	610	620	630	640	650	660	670	680	690	700	710	720	730	740	750
0.0	0.0	20.0	21.1	21.3	21.5	21.6	21.8	22.0	22.2	22.3	22.5	22.7	22.9	23.0	23.2	23.4	23.6	23.8
2.4	0.5	19.5	21.3	21.5	21.7	21.9	22.1	22.2	22.4	22.6	22.8	23.0	23.1	23.3	23.5	23.7	23.9	24.1
4.9	1.0	19.0	21.6	21.8	22.0	22.1	22.3	22.5	22.7	22.9	23.1	23.3	23.4	23.6	23.8	24.0	24.2	24.4
7.5	1.5	18.6	21.9	22.0	22.2	22.4	22.6	22.8	23.0	23.2	23.4	23.6	23.8	23.9	24.1	24.3	24.5	24.7
10.3	2.0	18.1	22.1	22.3	22.5	22.7	22.9	23.1	23.3	23.5	23.7	23.9	24.1	24.3	24.5	24.7	24.9	25.0
13.2	2.5	17.6	22.4	22.6	22.8	23.0	23.2	23.4	23.6	23.8	24.0	24.2	24.4	24.6	24.8	25.0	25.2	25.4
16.3	3.0	17.1	22.7	22.9	23.1	23.3	23.6	23.8	24.0	24.2	24.4	24.6	24.8	25.0	25.2	25.4	25.6	25.8
19.5	3.5	16.7	23.1	23.3	23.5	23.7	23.9	24.1	24.3	24.5	24.7	25.0	25.2	25.4	25.6	25.8	26.0	26.2
23.0	4.0	16.2	23.4	23.6	23.8	24.1	24.3	24.5	24.7	24.9	25.1	25.4	25.6	25.8	26.0	26.2	26.4	26.6
26.6	4.5	15.7	23.8	24.0	24.2	24.4	24.7	24.9	25.1	25.3	25.6	25.8	26.0	26.2	26.4	26.7	26.9	27.1
30.5	5.0	15.2	24.2	24.4	24.6	24.9	25.1	25.3	25.5	25.8	26.0	26.2	26.5	26.7	26.9	27.1	27.4	27.6
34.6	5.5	14.8	24.6	24.8	25.1	25.3	25.5	25.8	26.0	26.2	26.5	26.7	26.9	27.2	27.4	27.6	27.9	28.1
39.0	6.0	14.3	25.0	25.3	25.5	25.8	26.0	26.2	26.5	26.7	27.0	27.2	27.5	27.7	27.9	28.2	28.4	28.7
43.7	6.5	13.8	25.5	25.8	26.0	26.3	26.5	26.8	27.0	27.3	27.5	27.8	28.0	28.3	28.5	28.8	29.0	29.3
48.7	7.0	13.3	26.0	26.3	26.5	26.8	27.0	27.3	27.6	27.8	28.1	28.3	28.6	28.9	29.1	29.4	29.6	29.9
54.1	7.5	12.9	26.6	26.8	27.1	27.4	27.6	27.9	28.2	28.4	28.7	29.0	29.2	29.5	29.8	30.0	30.3	30.6
59.9	8.0	12.4	27.1	27.4	27.7	28.0	28.3	28.5	28.8	29.1	29.4	29.6	29.9	30.2	30.5	30.8	31.0	31.3
66.2	8.5	11.9	27.8	28.1	28.4	28.6	28.9	29.2	29.5	29.8	30.1	30.4	30.7	31.0	31.2	31.5	31.8	32.1
72.9	9.0	11.4	28.5	28.8	29.1	29.4	29.7	30.0	30.3	30.6	30.9	31.2	31.5	31.8	32.1	32.4	32.7	33.0
80.3	9.5	11.0	29.2	29.5	29.8	30.2	30.5	30.8	31.1	31.4	31.7	32.0	32.3	32.7	33.0	33.3	33.6	33.9
88.3	10.0	10.5	30.0	30.4	30.7	31.0	31.3	31.7	32.0	32.3	32.6	33.0	33.3	33.6	33.9	34.3	34.6	34.9
97.0	10.5	10.0	30.9	31.3	31.6	32.0	32.3	32.6	33.0	33.3	33.7	34.0	34.3	34.7	35.0	35.4	35.7	36.0
106.6	11.0	9.5	31.9	32.3	32.6	33.0	33.3	33.7	34.1	34.4	34.8	35.1	35.5	35.8	36.2	36.6	36.9	37.3
117.2	11.5	9.0	33.0	33.4	33.7	34.1	34.5	34.9	35.3	35.6	36.0	36.4	36.8	37.1	37.5	37.9	38.3	38.6
129.0	12.0	8.6	34.2	34.6	35.0	35.4	35.8	36.2	36.6	37.0	37.4	37.8	38.2	38.6	39.0	39.3	39.7	40.1
142.1	12.5	8.1	35.6	36.0	36.4	36.8	37.2	37.6	38.1	38.5	38.9	39.3	39.7	40.2	40.6	41.0	41.4	41.8
156.8	13.0	7.6	37.1	37.5	38.0	38.4	38.8	39.3	39.7	40.2	40.6	41.1	41.5	42.0	42.4	42.8	43.3	43.7
173.4	13.5	7.1	38.8	39.3	39.7	40.2	40.7	41.2	41.6	42.1	42.6	43.0	43.5	44.0	44.5	44.9	45.4	45.9
192.4	14.0	6.7	40.8	41.3	41.8	42.3	42.8	43.3	43.8	44.3	44.8	45.3	45.8	46.3	46.8	47.3	47.8	48.3
214.2	14.5	6.2	43.0	43.6	44.1	44.7	45.2	45.7	46.3	46.8	47.4	47.9	48.5	49.0	49.5	50.1	50.6	51.2
239.4	15.0	5.7	45.7	46.3	46.8	47.4	48.0	48.6	49.2	49.8	50.4	51.0	51.5	52.1	52.7	53.3	53.9	54.5
269.2	15.5	5.2	48.8	49.4	50.1	50.7	51.4	52.0	52.6	53.3	53.9	54.6	55.2	55.8	56.5	57.1	57.8	58.4
304.7	16.0	4.8	52.6	53.3	54.0	54.7	55.4	56.1	56.8	57.5	58.2	58.9	59.6	60.3	61.0	61.7	62.4	63.1

Higher Heating Value (HHV)	8,800 Btu/Lb		Carbon	50%
Moisture	30%		Hydrogen	6.5%
			Ultimate CO2	20.0%

600 - 750 F
WOOD 30% MOISTURE

FUEL SAVINGS

The purpose of this section is to show how to compute fuel savings based on change in efficiency.

$$\text{Fuel Savings}(\%) = \frac{\text{New Efficiency- As Found Efficiency}}{\text{New Efficiency}}$$

Example 1. The **As Found Efficiency** of a Natural Gas fired boiler is 73%, the boiler is **Tuned Up** to a new efficiency of 75%. What is the percent fuel savings?

Answer:

Example 2. The **Maximum Achievable Efficiency** for this boiler is about 81.7%. What will the % fuel savings be when this is achieved?

Answer:

Example 3. The **Maximum Attainable Efficiency** for this boiler is about 86.2%. What will the % fuel savings be when this is achieved?

Answer:

[Answers: **1.** 2.67%, **2.** 10.6%, **3.** 15.31%]

Boiler Plant Calculations

Maximum Economically Achievable Efficiency Levels

Fuel	Rated Capacity Million BTU's/HR		
	10 -16	16 - 100	100 - 250
Gas	80.1%	81.7%	84.0%
Oil	84.1%	86.7%	88.3%
Coal Stoker	81.6%	83.9%	85.5%
Pulverized	83.3%	86.8%	88.8%

Maximum Attainable Efficiency Levels

Fuel	Rated Capacity Million BTU's/HR		
	10 -16	16 - 100	100 - 250
Gas	85.6%	86.2%	86.5%
Oil	88.8%	89.4%	89.7%
Coal Stoker	86.4%	87.0%	87.3%
Pulverized	89.5%	90.1%	90.4%

REDUCING EXCESS AIR
TO
IMPROVE EFFICIENCY

EXAMPLE 1. A Natural Gas boiler has a stack temperature of 600F. What will the efficiency increase and fuel savings be if the average excess air is reduced from 90% to 15%   (CO_2 from 5.9% to 10.1%)?

As found efficiency	70.4%
Net stack temperature 600F-80F	520F
CO_2	5.9%
Efficiency with reduced excess air	77.4%
Net stack temperature 600F-80F	520F
CO_2	10.1%

Fuel savings _____%

Example 2. What will the fuel savings be with the same excess air reduction with a stack temperature of 300F instead of 600F?

As found efficiency	81.8%
Net stack temperature 300F-80F	220F
CO_2	5.9%
Efficiency with reduced excess air	84.7%
Net stack temperature 300F-80F	220F
CO_2 10.1%	

Fuel savings _____%

What conclusion can you draw from these examples?

[Answers: **1.** 6.4%, **2.** 2.0%, **3.** There is greater opportunity for fuel savings at the higher temperature 6.4% Vs 2.0%]

REDUCING STACK
TEMPERATURE

The purpose of this example is to demonstrate the loss of fuel dollars caused by the gradual fouling of heat exchange surfaces.

The stack temperature has risen from 480F to 680F since the boiler firesides were last cleaned. How much is this costing in terms of the monthly fuel bill?

Fuel   No. 2 Fuel Oil
Monthly fuel costs $150,000

Tuned Up Conditions:

    Flue gas oxygen   2%
    Stack temperature   480F
    Air Temperature   80F
    Net stack temperature   400F

    Tuned up Efficiency 86%

Present Conditions:

    Flue gas oxygen   2%
    Stack temperature   680F
    Air Temperature   80F
    Net stack temperature   600F

    Efficiency   82.1%

1. Percent efficiency loss   _____%

2. Percent fuel loss   _____%

3. Monthly dollar loss $_____

4. Annual dollar loss $_____

5. Would it be cost effective to clean the heat exchange surfaces?

[Answers: **1.** 3.9%, **2.** 4.75%, **3.** $7,125, **4.** $85,505, **5.** "It depends"]

Boiler Plant Calculations

Analyzing Costs and Benefits

ECONOMIZER VS OXYGEN TRIM

Part I: Economic Analysis, Estimating Percent Savings and Dollar Savings

As found conditions:

Fuel	Natural Gas
Annual Fuel Bill	$100,000
Oxygen O_2	11.0%
Net stack temperature	600°F
Efficiency	70.3%

Option 1. Install an Oxygen Trim system with an installed cost of $10,000 to reduce excess air to 2.0%.

New oxygen O_2 level	2%
Net stack temperature	600°F
New efficiency	79.6%

1. Efficiency Improvement _____ %

2. Percent Fuel Savings _____ %

3. Fuel Savings Dollars $_____/YR

Option 2. Install economizer with an installed cost of $30,000 to reduce stack temperature to 200°F.

Oxygen O2 level	11.0%
Net stack temperature	200°F
New efficiency	85.4%

4. Efficiency Improvement _____ %

5. Percent Fuel Savings _____ %

6. Fuel Savings Dollars $_____/YR

[Answers: **Option 1; 1.** 9.3%, **2.** 11.7%, **3.** $11,700/Yr. **Option 2; 4.** 15.1%, **5.** 17.68%, **6.** $17,681/Yr]

Analyzing Costs and Benefits

ECONOMIZER VS OXYGEN TRIM

Part II: Economic analysis, Payback Period

Payback Period $$PP = \frac{FC}{S - C}$$

PP = Payback period
FC = First cost
 S = Annual fuel savings
 C = Annual maintenance costs

Option 1. Calculate payback period for the oxygen trim system if the maintenance cost is $1,000/yr.

$$PP = \frac{10,000}{11,700 - 1,000} = \underline{\hspace{1.5in}} \text{ YR}$$

Option 2. Calculate payback period for an economizer if the maintenance cost is $1,000/yr.

$$PP = \frac{30,000}{17,681 - 1,000} = \underline{\hspace{1.5in}} \text{ YR}$$

[Answers: **Option 1.** .93 yr (11.2 Mo), **Option 2.** 1.8 Yr]

Boiler Plant Calculations

Analyzing Costs and Benefits

ECONOMIZER VS OXYGEN TRIM

Part III: Economic Analysis, Return on Investment

Return on investment $ROI (\%) = \dfrac{S\text{-}DC}{FC} \times 100$

  ROI = Return on Investment
   S = Annual fuel savings $/Yr
  DC = Depreciation charge   (First cost/Estimated Lifetime)
  FC = First Cost

Option 1.   Calculate the ROI for the oxygen trim system if the estimated lifetime is 15 years:

ROI (%):

$$\frac{11{,}700 - 667}{10{,}000} =$$

Option 2. Calculate the ROI for an economizer if the estimated lifetime is 15 years:

ROI (%):

$$\frac{17{,}681 - 2000}{30{,}000} =$$

[Answers: **Option 1.** 110%, **Option 2.** 52.3%]

196

Boiler Plant Calculations

Analyzing Costs and Benefits

ECONOMIZER VS OXYGEN TRIM

Part IV: Economic Analysis, Installing Oxygen Trim and Economizer Separately and Together.

Option 3. Assuming the oxygen trim system has been installed, what are the benefits from installing the economizer under the new operating conditions?

Efficiency with Oxygen Trim System	79.6%
Efficiency with economizer	86.8%
Annual Fuel Savings (%)	8.3%
Annual Fuel Savings ($)	$8,295/Yr

$$\text{Payback Period} = \frac{30,000}{8,295 - 1,000} = 4.1 \text{ Yr}$$

$$\text{Return on Investment} = \frac{8,295 - 2,000}{30,000} = 21\%$$

Option 4. What would the payback period and return on investment be if the oxygen trim and economizer were both installed as the same project?

As found efficiency	70.3%
Optimized efficiency: oxygen trim and economizer	86.8%
Optimized fuel savings (%)	19.0%
Optimized fuel savings ($)	$19,000/Yr

$$\text{Payback Period} = \frac{30,000 + 10,000}{19,000 - 2,000} = 2.35 \text{ Yr}$$

$$\text{Return on Investment} = \frac{19,000 - 2,667}{40,000} = 40\%$$

[Answers: **Option 3;** 4.1 yr, 21%; **Option 4;** 2.35 yr, 40%]

Boiler Plant Calculations

COST OF MONEY

The cost of money may affect energy saving projects. The purpose of this section is to illustrate how the longer payback periods are affected by the cost of money and also how interest rates change the time needed to recoup investments.

Your company must borrow money at various interest rates, how long will it take to recoup the investment required for energy conservation projects with different payback periods and for various interest rates?

    (1) Find the discount rate column. (p. 199)
    (2) Locate simple payback period under the assumed interest rate.
    (3) The approximate time to recoup investment will be found in the left column under the "lifetime" heading.

Note. These values are approximate and are given only to illustrate relative values)

	Payback Period	Discount Rate	Time to Recoup Investment
Option 1.	1.5 Yr	20%	_____
Option 2.	2.6 Yr	20%	_____
Option 3.	5.0 Yr	20%	_____
Option 4.	5.0 Yr	15%	_____
Option 5.	5.0 Yr	10%	_____
Option 6.	5.0 Yr	5%	_____

[Answers: **Option 1.** < 2 yr, **Option 2.** < 3 yr, **Option 3.** > 25 yr, **Option 4.** 10 yr, **Option 5.** 8 yr, **Option 6.** 6 yr]

PRESENT WORTH FACTORS (PWF)

Lifetime (EL)	Discount Rate (D)				
	5%	10%	15%	20%	25%
1	0.952	0.909	0.870	0.833	0.800
2	1.859	1.736	1.626	1.528	1.440
3	2.723	2.487	2.283	2.106	1.952
4	3.546	3.170	2.855	2.589	2.362
5	4.329	3.791	3.352	2.991	2.689
6	5.076	4.355	3.784	3.326	2.951
7	5.786	4.868	4.160	3.605	3.161
8	6.463	5.335	4.487	3.837	3.329
9	7.108	5.759	4.772	4.031	3.463
10	7.722	6.145	5.019	4.192	3.571
11	8.306	6.495	5.234	4.327	3.656
12	8.863	6.814	5.421	4.439	3.725
13	9.394	7.103	5.583	4.533	3.780
14	9.899	7.367	5.724	4.611	3.824
15	10.380	7.606	5.847	4.675	3.859
16	10.838	7.824	5.954	4.730	3.887
17	11.274	8.022	6.047	4.775	3.910
18	11.690	8.201	6.128	4.812	3.928
19	12.085	8.365	6.198	4.843	3.942
20	12.462	8.514	6.259	4.870	3.954
21	12.821	8.649	6.312	4.891	3.963
22	13.163	8.772	6.359	4.909	3.970
23	13.489	8.883	6.399	4.925	3.976
24	13.799	8.985	6.434	4.937	3.981
25	14.094	9.077	6.464	4.948	3.985

The above table is calculated from the following equation:

$$PWF = \frac{1 - (1 + D)^{-EL}}{D}$$

where D is discount rate expressed as a fraction and EL is the expected lifetime of the project in years.

Boiler Plant Calculations

Heat Losses in the Condensate Return System

The purpose of this section is to illustrate how energy is lost in the condensate recovery system and by live steam lost from the distribution system.

Example 1. Assuming that condensate should return to the receiver at 194°F, what losses are involved for each pound of cold make-up water involved?

City Water	50°F	18 Btu/lb
Hot condensate	194°F	162 Btu/lb
Water heating		144 Btu/lb
Water heating including boiler losses (80% Eff)		180 Btu/lb
100 PSI Steam		1189 Btu/lb

$$\textbf{Fuel Loss} \quad = \quad \frac{\textbf{180 Btu/lb}}{\textbf{1189 Btu/lb}} \quad = \quad \textbf{15\%}$$

Example 2. What is the total loss involved in steam leaks?

City Water	50°F	18 Btu/lb
100 PSI Steam		1189 Btu/lb
Steam energy including boiler losses (80% Eff)		1464 Btu/lb

$$\textbf{Fuel Loss} \quad = \quad \frac{\textbf{1464 Btu/lb}}{\textbf{1189 Btu/lb}} \quad = \quad \textbf{123\%}$$

Example 3. Assume the condensate should be returning to the plant at 194°F, it has been returning at 120°F instead. What losses are involved?

Hot condensate	194°F	162 Btu/lb
Actual condensate	120°F	88 Btu/lb
Energy loss		74 Btu/lb
Fuel energy loss including boiler (80% Eff)		93 Btu/lb

$$\text{Fuel Loss} = \frac{93 \text{ Btu/lb}}{1189 \text{ Btu/lb}} = 7.8\%$$

Example 4. Derive the "Rule of Thumb" relationship about how condensate temperature influences boiler efficiency. How many degrees must the feed-water or condensate change to cause approximately a one percent change in boiler system efficiency?

100 PSI steam	1189 Btu/lb
Water temperature rise of 11 degrees F	11 Btu/lb
Actual boiler energy required for 11°F rise (80% Eff)	13.8 Btu/lb

$$\frac{13.75 \text{ Btu/lb}}{1189 \text{ Btu/lb}} = 1.16\%$$

Rule of Thumb: "A 1% increase in efficiency occurs with each 11°F rise in condensate return or feed-water input temperature".

Boiler Plant Calculations

Gage Pressure (psig)	Saturation or Boiling Temperature (Degrees F)	Specific Volume (Cu. Ft./Lb.)	Heat Content Above 32 Degrees F		
			Sensible Heat or Heat of Liquid (Btu/lb.)	Latent Heat or Heat of Evaporation (Btu/lb.)	Total Heat (Btu/lb.)
0	212.0	26.80	180.1	970.3	1150.4
1	215.5	25.13	183.6	968.1	1151.7
2	218.7	23.72	186.8	966.0	1152.8
3	221.7	22.47	189.8	964.1	1153.9
4	224.5	21.35	192.7	962.3	1155.0
5	227.3	20.34	195.5	960.5	1156.1
6	229.9	19.42	198.2	958.8	1157.0
7	232.4	18.58	200.7	957.2	1157.9
8	234.9	17.81	203.2	955.6	1158.8
9	237.2	17.11	205.6	954.1	1159.7
10	239.5	16.46	207.9	952.5	1160.4
11	241.7	15.86	210.1	951.1	1161.2
12	243.8	15.31	212.2	949.7	1161.9
13	245.9	14.79	214.3	948.3	1162.6
14	247.9	14.31	216.4	946.9	1163.2
15	249.8	13.86	218.3	945.6	1163.9
16	251.7	13.43	220.3	944.3	1164.6
17	253.6	13.03	222.2	943.0	1165.2
18	255.4	12.66	224.0	941.8	1165.8
19	257.1	12.31	225.7	940.6	1166.3
20	258.8	11.98	227.5	939.5	1167.0
21	260.5	11.67	229.2	938.3	1167.5
22	262.2	11.37	230.9	937.2	1168.1
23	263.8	11.08	232.5	936.1	1168.6
24	265.4	10.82	234.1	935.0	1169.1
25	266.9	10.56	235.6	934.0	1169.6
30	274.1	9.45	243.0	928.9	1171.9
35	280.7	8.56	249.8	924.2	1174.0
40	286.8	7.82	256.0	919.8	1175.8
45	292.4	7.20	261.8	915.7	1177.5
50	297.7	6.68	267.2	911.8	1179.0
55	302.7	6.23	272.4	908.1	1180.5
60	307.3	5.83	277.2	904.6	1181.8
65	311.8	5.49	281.8	901.3	1183.1
70	316.4	5.18	286.2	898.0	1184.2
75	320.1	4.91	290.4	894.8	1185.2
80	323.9	4.66	294.4	891.9	1186.3
85	327.6	4.44	298.2	899.0	1187.2
90	331.2	4.24	301.9	886.1	1188.0
95	334.6	4.06	305.5	883.3	1188.8
100	337.9	3.89	308.9	880.7	1189.6
105	341.1	3.74	312.3	878.1	1190.4
110	344.2	3.59	315.5	875.5	1191.0
115	347.2	3.46	318.7	873.0	1191.7
120	350.1	3.34	321.7	870.7	1192.4

"Steam Tables", heat content of steam and water at various pressures.

Boiler Plant Calculations

Gage Pressure (psig)	Saturation or Boiling Temperature (Degrees F)	Specific Volume (Cu. Ft./Lb.)	Heat Content Above 32 Degrees F		
			Sensible Heat or Heat of Liquid (Btu/lb.)	Latent Heat or Heat of Evaporation (Btu/lb.)	Total Heat (Btu/lb.)
125	352.9	3.23	324.7	868.3	1193.0
130	355.6	3.12	327.6	865.9	1193.5
135	358.3	3.02	330.4	863.7	1194.1
140	360.9	2.93	333.1	861.5	1194.6
145	363.4	2.84	335.8	859.3	1195.1
150	365.9	2.76	338.4	857.2	1195.6
155	368.3	2.68	340.9	855.0	1195.9
160	370.6	2.61	343.4	853.0	1196.4
165	372.9	2.54	345.9	850.9	1196.8
170	375.2	2.47	348.3	848.9	1197.2
175	377.4	2.41	350.7	846.9	1197.6
180	379.5	2.35	353.0	845.0	1198.0
185	381.6	2.30	355.2	843.1	1198.3
190	383.7	2.24	357.4	841.2	1198.6
195	385.8	2.19	359.6	839.2	1198.8
200	387.8	2.13	361.9	837.4	1199.3
210	391.7	2.04	366.0	833.8	1199.9
220	395.5	1.95	370.1	830.3	1200.4
230	399.1	1.88	374.1	826.8	1200.9
240	402.7	1.81	377.8	823.4	1201.3
250	406.1	1.74	381.6	820.1	1201.7
260	409.4	1.68	385.2	816.9	1202.1
270	412.6	1.62	388.7	813.7	1202.4
280	415.7	1.56	392.1	810.5	1202.7
290	418.8	1.52	395.5	807.5	1202.9
300	421.8	1.47	398.7	804.5	1203.2
400	448.2	1.12	428.1	776.4	1204.6
500	470.0	0.90	452.9	751.3	1204.3
600	488.8	0.75	474.6	728.3	1202.9

"Steam Tables", heat content of steam and water at various pressures (continued).

Boiler Plant Calculations

Distribution System Losses

The purpose of this section is to illustrate the major losses in steam distribution systems.

Steam system load	20,000 lb/hr
Steam system average Input	22.5 MBtu/hr
Net energy in steam	1125 Btu/lb
Distribution system pressure	125 PSIG
Condensate system pressure	Atmospheric
Condensate return temperature	120°F
Make up water temperature	60°F

1. What is the flash loss?

Heat in the condensate on the high pressure side   325 Btu/hr
of the steam trap.

Heat in condensate atmospheric pressure.   180 Btu/lb
Latent heat of flash steam   970 Btu/lb

$$\text{Flash} = \frac{325-180}{970} = 15\%$$

Energy lost with flash steam:

15% X 20,000 lb/hr X 970 Btu/lb = 2.91 million Btu/hr

Energy loss   = 2.91/22.5   =   **12.93% Loss**

2. Condensate system losses.
Condensate reaches the receiver at 120F, having cooled from 212F losing 92 Btu/lb for a total of 1.53 MBtu/hr. (83% of the condensate returns and 17% is lost in the system)

83% X 20,000 lb/hr X 92 Btu/lb = 1.53 Million Btu/hr

Energy loss   1.53/22.5   =   **6.8% Loss**

3. Cold Make-up water. Because of steam and condensate leaks, 17% make-up feed water at 60F is required.

Energy loss = 17% X 20,000 lb/hr X 152 Btu/lb   =   0.52 Million Btu/hr

Energy Loss .52/22.5   =   **2.3% Loss**

4. Average trap leakage is 10% live steam in the system.

Energy loss = .1 X 20,000 PPH X 1125 Btu/lb   = 2.25 Million Btu/hr

Loss   2.25/22.5   =   **10% Loss**

Total System losses   **32.03%**

Steam system efficiency   **67.97%**

Boiler Plant Calculations

FLASH STEAM TABLE

INITIAL PRESSURE PSIG	TEMP OF LIQUID	% FLASH AT ATMOSPHERIC PRESSURE	PERCENT OF STEAM FLASHED AT REDUCED PRE				
			5 PSI	10 PSI	15 PSI	20 PSI	25 PSI
100	338	13.0	11.5	10.3	9.3	8.4	7.6
125	353	14.5	13.3	11.8	10.9	10.0	9.2
150	366	16.0	14.6	13.2	12.3	11.4	10.6
175	377	17.0	15.8	14.4	13.4	12.5	11.6
200	388	18.0	16.9	15.5	14.6	13.7	12.9
225	397	19.0	17.8	16.5	15.5	14.7	13.9
250	406	20.0	18.8	17.4	16.5	15.6	14.9
300	421	21.5	20.3	19.0	18.0	17.2	16.5
350	435	23.0	21.8	20.5	19.5	18.7	18.0
400	448	24.0	23.0	21.8	21.0	20.0	19.3
450	459	25.0	24.3	23.0	22.0	21.3	20.0
500	470	26.5	25.4	24.1	23.2	22.4	21.7
550	480	27.5	26.5	25.2	24.3	23.5	22.8
600	488	28.0	27.3	26.0	25.0	24.3	23.6
BTU PER POUND OF FLASH STEAM		1150	1155	1160	1164	1167	1169
DEG F OF FLASH STEAM & LIQUID		212	225	240	250	259	267
STEAM VOLUME CUFT/LB		26.8	21.0	16.3	13.7	11.9	10.5

206

Boiler Plant Calculations

Boiler Blowdown
Heat Recovery

The purpose of this section is to demonstrate the potential dollar savings available from managing the boiler water system.

1. Flash Steam Recovery

What percentage of blowdown will flash to steam at 5 psig from the blowdown of a boiler operating at 150 psig?

$$\% \text{ Steam Flashed} = \frac{h_1 - h_2}{V_1}$$

h_1 Heat in water leaving boiler.
h_2 Heat in water at flash pressure.
V_1 Latent heat of steam at flash pressure.

Steam Table Information

Boiler pressure 150 psig $h_1 = $ _____   [Answer: 338.4 Btu/lb]
Flash tank pressure 5 psig $h_2 = $ _____   [Answer: 195.5 Btu/lb]
Latent heat of steam $v_1 = $ _____   [Answer: 960.5 Btu/lb]
at 5 psig

% Steam Flashed = _____   [Answer: 15%]

2. How many Btu/day will be recovered from the flash tank if the daily blowdown is 80,000 lb/day?

Btu/day = % steam flashed x blowdown lb/day x (Btu/lb @ 5 psi Steam) =

(At 5 psig the energy in the flash steam is 1156 Btu/lb)

Btu/day = ...

[Answer: Btu/day = .15 X 80,000 X 1156 = 13.9 Million Btu/day]

3. If you are firing a boiler with no. 2 oil with 140,000 btu/gal which costs $1.00 per gallon, how much money can be saved in a day?

$$\text{\$ saved/day} = \frac{\text{Btu/day X cost/gal}}{\text{Btu/gal x E}} =$$

E = Boiler efficiency = 75%

Dollar savings = ...

[Answer: 13.9 Million Btu/day X $1.00 / 140,000 X .75 = $132.90]

4. Heat exchanger recovery

The heat exchanger now reduces the blowdown water temperature to 90F, how many Btu's a day can be recovered from this water?

Btu/day recovered = Blowdown (lb/day) X 1- % flashed X h_2-h_4 =

Heat in water at flash pressure (5 psig)..............h_2 = _____[Answer: 195.5 Btu/lb]
Heat in water at heat exchanger outlet (90F)..........h_4 = _____[Answer: 59 Btu/lb]

Btu/day recovered = ... = _____ [Answer: 9.28 Million Btu/day]

$$\text{Dollar savings/day} = \frac{\text{Btu/day X \$1.00}}{140,000 \text{ Btu/gal X .75}} = _____$$ [Answer: 88.4 Million Btu/day]

5. Total value of blowdown heat recovered:

Flash steam.....................................$_____/day [Answer: $132.90/day]

Heat Exchanger................................$_____/day [Answer: $88.40/day]

         Total $_____/day [Answer: $221.30/day]

        Annual Savings $_____/Yr [Answer: $80,774.5/Yr]

Boiler Plant Calculations

SAVING ENERGY
BY
IMPROVING CONDENSATE RETURN

In this case there is no blowdown heat recovery system and the challenge is to conserve energy by reducing the blowdown as much as possible. There are several ways to do this such as repairing system leaks, improve feedwater quality and increasing cycles of concentration.

Calculate the savings possible in a plant exporting 2 million pounds of steam a day.

	As Found Condition	Optimized Condition
Condensate return	20%	80%
Makeup water	80%	20%
Blowdown	14% 280,000 lb/day	8% 160,000 lb/day

$$\text{\$ Savings} = \frac{BR \times H \times C}{V \times \%E}$$

BR = blowdown reduction lb
H = Heat content of blowdown (100 psi)
$\quad$ H = h_1 - h_5
$\quad\quad$ h_1-heat in water leaving boiler as blowdown
$\quad\quad$ h_5-heat in makeup feed water

$= h_1$-h_5 (309 - 28) = 281 Btu/lb
$\quad$ C = Cost of fuel \$1.00/gal
$\quad$ V = Heating Value of Fuel (No.2 fuel- 140,000 Btu/lb)
$\quad$ E = Boiler Efficiency - 80%

$$\text{\$ Savings} = \frac{120,000 \text{ lb/day} \times 281 \text{ Btu/lb} \times \$1.00\text{/gal}}{140,000 \text{ Btu/gal} \times .80}$$

\$ Savings $\quad$ = $\quad$ \$301 /day
\$ Savings $\quad$ = $\quad$ \$109,865 /Yr

REDUCING BLOWDOWN LOSSES
BY
INCREASE CYCLES OF CONCENTRATION

Example 1. A 300 psi boiler rated at 250,000 pounds of steam per hour uses 50% make-up water averaging 1,875,000 pounds a day. It is anticipated that the cycles of concentration can be raised from 7 to 14. What will the savings be?

$$\text{Formula} \quad A \quad = \quad \frac{C2-C1}{(C1 \times C2) - C1}$$

C1 = Present cycles of concentration   = 7

C2 = Anticipated cycles of concentration = 14

The chart indicates a savings of 2.85 Million Btu per 100,000 lb of makeup reduction.

With a cost of steam of $8.00 per million Btus, estimate the dollar savings.

Savings = 18.75 X $8.00 X 2.85 = $427.50/day

= $156,037/Yr

Boiler Plant Calculations

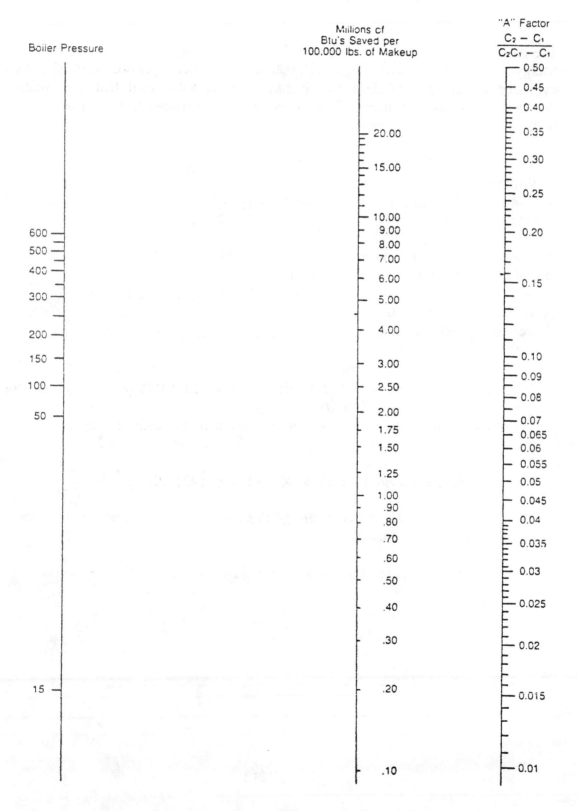

BTU'S SAVED BY INCREASING
CYCLES OF CONCENTRATION

Boiler Pressure

Millions of
Btu's Saved per
100,000 lbs. of Makeup

"A" Factor
$$\frac{C_2 - C_1}{C_2C_1 - C_1}$$

ELECTRICAL LOSSES

The purpose of this section is to show how electrical losses play a part in the overall energy consumption of a boiler plant. Operating electrical equipment, especially if it is oversized for the load on the plant can add to unnecessary plant losses. The examples in this section are based on nameplate data. Field measurements of electrial loads should be taken to confirm actual conditions.

Example 1. A 50,000 pound per hour (PPH) boiler never operates above 12,000 pounds per hour and is equippped with a 40 hp blower. A 12,000 pound per hour boiler can operate with a 5 hp blower.

With a cost for electrical power of 10 cents per kilowatt hour, what will the difference in cost be to run with the smaller blower?

(This example is for illustrative purposes only, actual electrical load will depend on the blower configuration and motor efficiency)

 A. How much does it cost to operate the 40 hp blower per hour?

 Ans:.......**$3.50/hr** (Approx)

 B. How much does it cost to operate the 5 hp blower per hour?

 Ans:.......**$0.45/hr** (Approx)

 C. Assuming 8,000 hours of operation a year, how much more will it cost to operate the 40hp blower than the 5hp blower?

 Ans:.........**$3.05 X 8,000 hr = $24,400/YR**

Example 2. A plant keeps a 100 hp feed pump on the line continuously where a 30 hp pump could be used. How much can be saved in a 8,000 hr year by installing a 30 hp pump?

 A. How much does it cost to run the 100 hp pump per year?

   Ans:.........$10.00/hr X 8,000 hr =$80,000/yr

 B. How much does it cost to run the 30 hp pump per Year?

   Ans:.......$2.75 X 8,000 hr = $22,000/yr

 C. What will the annual savings be if the 100 hp pump is replaced by a 30 hp pump?

   Ans:........$58,000/Yr (approx)

Results similar to the above examples can be achieved by use of **Variable Speed Drive** controllers.

Boiler Plant Calculations

HOURLY ELECTRICAL OPERATING COST

The dollar per hour ($/hr.) electrical operating cost must be considered in the boiler room operating cost. However, when compared to the $/hr. fuel cost, the $/hr. electrical cost is extremely small.

This again points out the importance of overall "fuel to steam" or boiler efficiency and affect on boiler room operating cost.

INSTRUCTIONS

1. Obtain the motor hp for the applicable boiler size (Bhp) from the Dimension and Rating Section.
2. Determine the ¢/Kw-hr. rate from the electric company.

3. Draw a line from the ¢/Kw-hr. rate to the applicable motor hp on Chart 50 or 51. Read the $/hr. electrical operating cost where the line intersects the vertical middle line.

HOURLY ELECTRICAL OPERATING COST

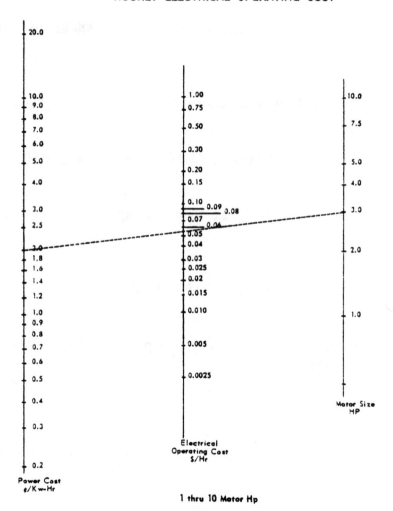

1 thru 10 Motor Hp

HOURLY ELECTRICAL OPERATING COST

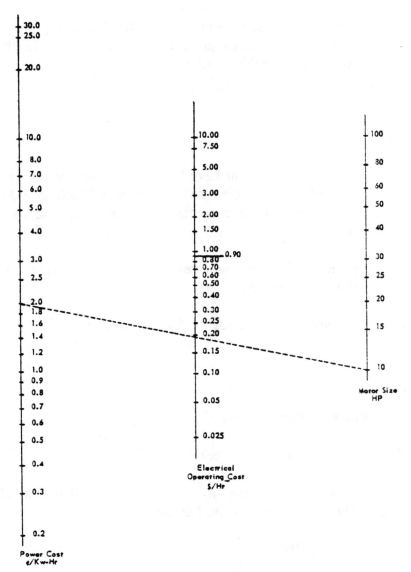

10 thru 100 Motor Hp

Boiler Plant Calculations

Setting The Fuel Oil Heater For
Proper Fuel Oil Viscosity
At The Burner

When the fuel oil temperature is not set to deliver oil at the proper viscosity, the burner can smoke and foul heat exchange surfaces. Also, higher levels of excess air will be needed to compensate for this problem. If the fuel oil heater temperature is too high the flame is affected also, becoming ragged with sparklers.

Situation:

As an economy measure, you have been given some heavy oil to burn which has a viscosity rating of 4,000 Sabolt Universal Seconds (SUS) at 100 degrees F. You notice that you have to use a lot more excess air with this fuel and you have to use your soot blowers more often.

Your burner technical manual indicates that your burner requires a viscosity of about 150 SUS for proper atomization.

What temperature is required for the fuel oil heater?

Solution:

Use the **Viscosity-Temperature** Chart.

1. Find 4,000 SUS at 100 degrees F.

2. Move down a line parallel to the No. 6 fuel oil line to 150 SUS and read the corresponding temperature for the fuel oil.

    Answer: 220°F

Note: If there is a significant piping run between the heater and the burner the actual temperature at the burner may be lower and the heater may have to be set for a higher temperature.

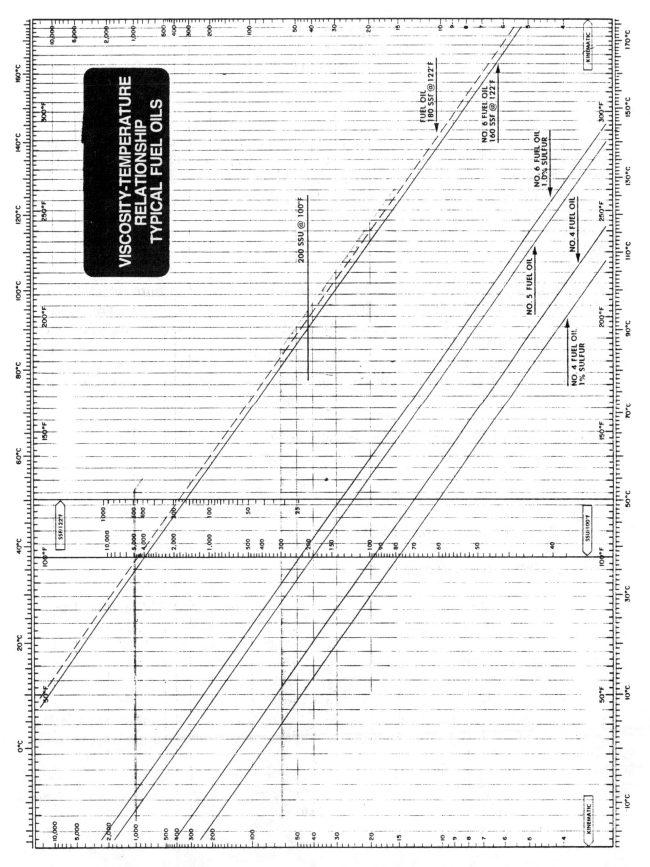

VISCOSITY-TEMPERATURE
RELATIONSHIP
TYPICAL FUEL OILS

Chapter 12

Waste Heat Recovery

The value of waste heat comes from the fact that it supplants additional input energy, reducing overall energy costs.

Waste Heat Recovery Opportunities

Flue gasses from a boiler represent a 17% to 30% (plus) opportunity for savings investments. This chapter will cover the technology for waste heat recovery including practical approaches to boiler efficiency improvement and other concepts for utilizing the recovered energy from flue gasses.

On the average the temperature of flue gasses leaving boilers is about 400°F ranging between 350°F and 650°F. Flue gas leaves the boiler at a temperature higher than the steam temperature for heat transfer to take place.

The boiler exhaust approach temperature, in general, will rise from 40°F to 150°F from low load to maximum (**Figure 12.1**). Many tests have shown these numbers to vary widely, so each boiler should be tested for its characteristic exhaust temperature, preferably after a cleaning and tune up, to establish the ideal temperatures for that particular boiler.

Where Waste Heat Can Be Used

A suitable use for waste energy is critical to any waste heat recovery project because it doesn't matter how much energy you can recover, the only thing that is going to save you money is to actually use the energy in your facility, and by doing so, decrease the amount of outside energy you have to purchase.

Typical uses for waste heat energy are:

a. Boiler feedwater heating
b. Makeup water heating
c. Combustion air preheating
d. Process heating
e. Domestic hot water
f. Generating electricity

If waste heat can be utilized in the boiler itself, a considerable advantage is gained by the fact that it is a self-controlling process requiring simple or no controls to regulate its application. If this same energy were to be used in a plant or building, it would be supplying a demand which would vary from the typical boiler operation and need additional controls. There might be periods when the energy wouldn't be needed thereby wasting it. Using the boiler for waste heat recovery provides an uninterruptable use of this energy.

Acid Formation a Limiting Factor in Waste Heat Recovery

One of the most important factors influencing stack gas heat recovery is the corrosion problem accompanying the cooling of the gas. Because the sulfuric acid dewpoint is higher than the water vapor dewpoint, heat recovery efforts must eventually contend with the acid dew point problem (**Figure 12.2**).

The acid dewpoint is that temperature at which acid begins to form (**Figure 12.3**). This temperature varies with the sulfur content of the fuel (**Figure 12.4**). To avoid the corrosive effects of acids the traditional

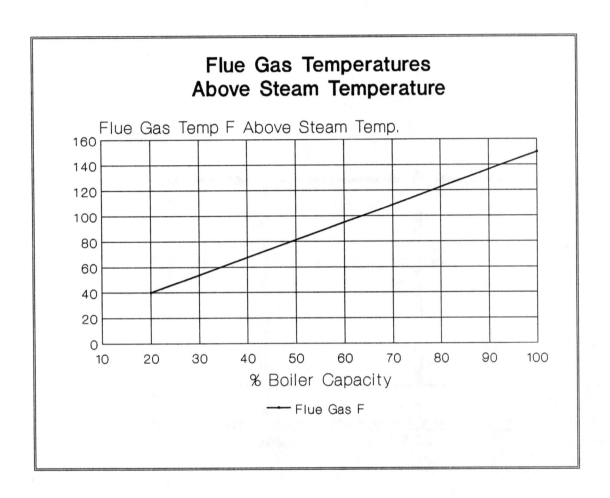

Figure 12.1. Typical exit gas temperature above steam temperature.

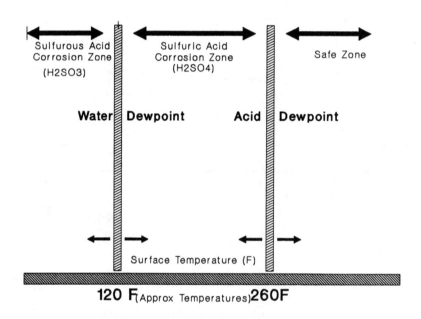

Figure 12.2. The relationship of acid dewpoint and water dewpoint to the formation of acids from the sulfur in fuel.

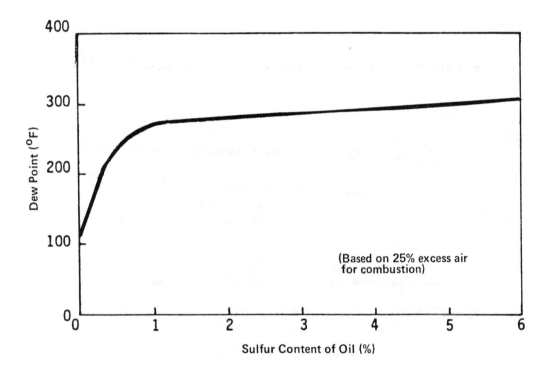

Figure 12.3. The relationship of acid dewpoint to the sulfur content of fuel.

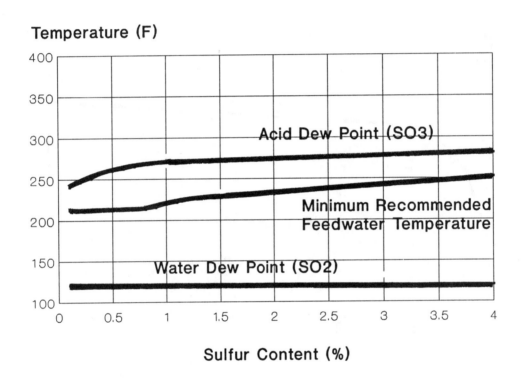

Figure 12.4. Minimum recommended feedwater temperature to avoid economizer tube corrosion.

practice has been to limit the heat recovery to a minimum stack gas exit temperature of 350°F. Within the past decade, however, corrosion-resistant materials have been developed and applied to heat recovery systems, many of which operate below the acid dewpoint.

What is Flue Gas

Before we go on, let's explore some basics in the field of waste heat recovery. The first applies to the nature of flue gasses which contains both dry gas and moisture. The amount of moisture in flue gasses varies with the amount of hydrogen in the fuel, each pound of hydrogen in the fuel combines with oxygen to form approximately 9 pounds of water (**Table 12.1**). This water is in the superheated state containing more than half the energy in the flue gas stream. As each fuel has a different carbon to hydrogen ratio the energy in the moisture in the flue gas stream will vary from fuel to fuel.

Species	Percentage Weight by Species for Fuel Types		
	Natural Gas	No. 2 Oil	Coal
Carbon	74.7	87.0	75.1
Hydrogen	23.3	12.5	4.8
Higher Heating (BTU/LB)	22,904	19,520	13,380

Table 12.1, percentage weight by species and higher heating values for fuel types.

Carbon as Fuel

The dry gas in the flue gas is formed from the combustion of carbon to carbon dioxide plus nitrogen which does not take part in the combustion process and any excess air above and beyond the amount needed for combustion.

Dry Gas and Moisture Losses

The carbon and hydrogen composition of various fuels determines the dry gas and moisture losses. **Table 12.2** shows how this affects flue gas losses and the minimum stack losses that cam be expected.

Fuel	Min. Dry Gas loss Loss (%)	Min. Moisture Loss (%)	Min. Stack Gas Loss (%)
Natural Gas	2.9	10.1	13.0
No. 2 Oil	5.1	6.4	11.5
No. 6 Oil	6.6	6.2	12.8
Coal	5.5	10.0	15.5

Table 12.2, flue gas losses due to dry gas and moisture.

Sensible and Latent Heat

Sensible heat is that heat which can be sensed with a thermometer or other temperature sensing instrument.

The formula for sensible heat transfer is:

$$Q = M \, C_p \, (T_2 - T_1)$$

$$Q = \text{Heat content BTU/HR}$$

223

Waste Heat Recovery

M = Flow rate LB/HR
C_p = Specific heat BTU/LB/F
T_1 = Stream temperature
T_2 = Reference temperature

Latent heat is the heat required for phase change, i.e. to change a liquid to a vapor (water to steam) or vice versa, without a change in temperature.

The formula for latent heat transfer is:

$$Q = M H$$

Q_1 = Heat content BTU/HR
M = Flow rate LB/HR
H = Heat of vaporization BTU/LB

Fuel

Combustion products from burning fuels with higher hydrogen content contain more water vapor and larger amounts of latent heat loss potential. Gas fired boilers are inherently less efficient than heavy oil fired units and represent better candidates for heat recovery. The type of fuel will also affect the maintainability and service life of a heat recovery system.

Natural gas is a clean-burning fuel and causes minimal corrosion problems in heat recovery hardware.

Fuel oil contains varying amounts of sulfur, which leads to acid corrosion problems.

Regenerators and Recouperators

Different terminologies have developed over the years in different industries referring to heat recovery process. The term regenerators has come to refer to the alternate heating and cooling of a media, such as plates or brickwork or other heat absorbing material with hot exhaust gasses, and then recapturing the heat by warming combustion or process air over the same media by manipulating gas and air streams.

Recuperators refers to the continuous operating (static) type or heat recovery unit using an intermediate wall between the hot and cold streams.

In boiler plants it is more common to hear the heat recovery apparatus referred to by name, (i.e. air preheater, economizer, etc.) than regenerator or recuperator.

Conventional Economizer

An economizer performs two functions, it reduces stack temperature and also heats boiler feed water (**Figure 12.5**).

A practical rule of thumb is that for every 40°F the stack temperature is reduced there is a corresponding 1% efficiency increase. On the water side, an increase of approximately 1% in efficiency is expected for each 11°F rise in feedwater temperature.

Economizers have been in use for a long time and it has been found that boilers operating at pressures of 75 psig or greater are excellent applications. One of the strongest points for installing an economizer is its compact size compared to other options.

Some general guidelines for economizer installations are:

    a. Average stack gas temperature of 450°F

    b. More than 2,500 operating hours

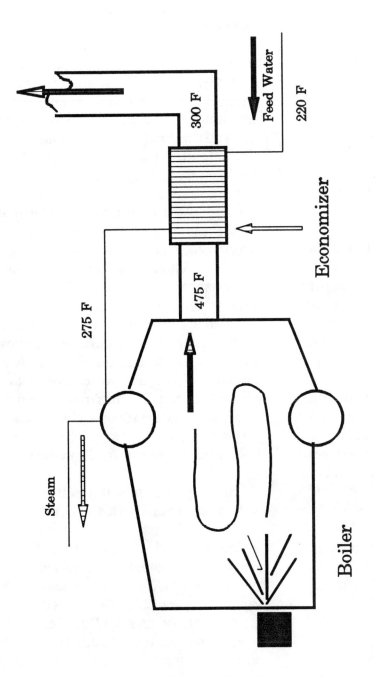

Figure 12.5, Economizer used to reduce stack temperature and raise feedwater temperature.

a year.

c. Stack flow rate more than 15,000 LB/HR

The maximum benefit that can be safely achieved is governed by a number of technical and physical limitations.

a. Economic considerations. Many economizers have paid for themselves in two years or less, the economizer then continues to return dividends from reduced energy costs long after it has been paid for.

b. When limited fuel availability or steam production threatens plant production an economizer can increase boiler capacity from 4% to 10%. If there is a need for more steam capacity, this may be the most cost effective way to do it rather than invest in a new boiler.

c. By the law of diminishing returns, an increase in heating surface does not provide, in equal proportion, for an increase in fuel savings. The flue gas temperature can not be reduced below the temperature of the incoming feed water or the acid formation temperature.

d. It may be possible that a controlling limitation may be imposed by the space available for the installation of an economizer.

e. When a boiler operates with a stack for induced natural draft on a negative pressure furnace, a limitation is imposed on the furnace draft due to the cooler exhaust gasses caused by the economizer. In this case an Induced draft fan may be required. In other cases of balanced draft or forced draft systems, the additional boiler efficiency can offset the additional draft requirement caused by the lower stack temperatures and pressure drop across the economizer.

f. Outlet water temperature is a limitation to prevent steam formation and water hammer in the economizer. An approach temperature of 40°F is customary for variable boiler load conditions. The outlet water temperature may be the dominant limitation in selecting economizer size.

g. Acid formation and condensation on the gas side of the heating surface is determined solely by the temperature of the surface which is essentially the same as the water temperature.

h. A minimum gas temperature of 250°F in the stack is desirable. Basically, this is to assure that the flue gas will be sufficiently buoyant to escape into the atmosphere and not mushroom around the stack and cause smoke or acid rain nuisance.

Cold-End Corrosion in Economizers

The major portion of sulfur in fuel is burned and forms sulfur dioxide (SO_2) in the flue gas; a small portion, 3 to 5 percent, is further oxidized to sulfur trioxide (SO_3). These oxides combine with moisture to form sulfurous (H_2SO_3) and sulfuric acid (H_2SO_4) vapors (**Figure 12.2**). When in contact with a surface below the Acid Dew Point (ADP), condensation takes place. The ADP is directly related to the amount of sulfur in the fuel as shown in **Figure 12.3**.

Because of the higher heat transfer coefficient in liquid-metal than in gas-metal heat transfer, the gas side metal temperature of an economizer is closer to the water temperature than the gas side temperature.

Design parameters for a conventional economizer are shown in **Table 12.3**, illustrating the minimum recommended feedwater temperature and flue gas exit temperatures.

Fuel Type	Minimum Inlet Water Temperature (°F)	Maximum Exit Flue Gas Temperature (°F)	Fin Density (Fins/In)
Natural Gas	210	300	5
No. 2 Fuel Oil	220	325	4
No 5 & 6 Fuel Oil	240	350	3
Coal	240	350	2

Table 12.3, typical design parameters for a conventional economizer.

Controlling Acid Formation

Controlling Economizer Inlet Temperature.

The most efficient and effective means of controlling economizer metal temperature is with a feedwater preheat system. The system is illustrated in **Figure 12.6**. It is essentially a feedwater preheater with sensors controlling the steam admission valves to the heater.

The steam admission valve sensors measure the water temperature entering the economizer and the temperature of the stack metal. Both of these surfaces are subject to corrosion from temperature exposure below the ADP. Using the heater insures neither surface will cool to that point. As an additional protection, the portion of the stack exposed to very cold weather conditions could be insulated to keep the metal temperature from becoming too cold, approaching the ADP.

Reduce excess air

Reducing excess air raises the dewpoint temperature in the stack gas (**Figure 12.7**). Research has shown a direct relationship between excess air and the formation of sulfur trioxide.

Use corrosion-resistant materials or sleeves

Corrosion-resistant alloy steels can be used in preventing corrosion, however their high cost normally prohibits their actual use. Interlocking cast iron sleeves over carbon steel tubes can also be used where severe acid conditions are anticipated. The cast iron sleeve will corrode, but it can be replaced at normal maintenance intervals. The cast iron sleeves protect the carbon steel tubes from contact with corrosive gases.

Improve fluid flow arrangement

Using parallel-flow tube arrangements rather than counter-flow arrangements increases economizer skin temperature. With the parallel flow arrangement, cold feedwater enters where the stack gas is the hottest, thus raising the tube surface temperature slightly.

Modulate feedwater flow through the economizer

The feedwater flow rate through the

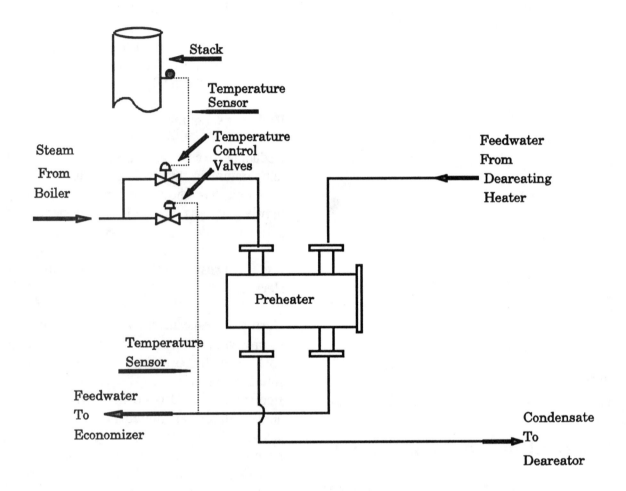

Figure 12.6. Feedwater preheater used to control cold end corrosion in the economizer and acid dew point in the stack. The steam energy used for heating is not lost as it recycles back into the boiler as heat in the feedwater.

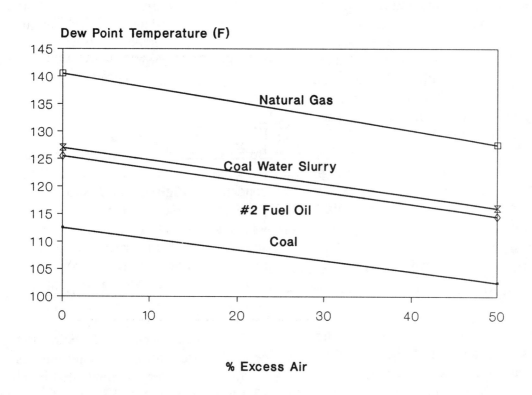

Figure 12.7. Relationship between initial water vapor dewpoint of flue gasses and the excess air level.

economizer can be controlled by diverting feedwater flow around the economizer during periods of low flue gas temperatures (**Figure 12.8**).

Insulate Stack Metal

Just as with metal temperatures in the economizer, flue gas temperatures will not necessarily determine the severity of acid corrosion, the same is true for the stack. It is the temperature of the metal, or stack wall, that is in contact with the flue gas that will determine the extent of acid corrosion. Stack insulation will aid in keeping stack temperatures above the ADP.

Alternatives to insulation are high temperature corrosion-resistant stack material such as Cor-Ten or Fiberglass reinforced plastic.

Condensing the Moisture in Flue Gasses

The major limitation to increased boiler efficiencies is the amount of energy tied up in latent heat. Until recent years equipment has not been available to capture and use this large source of waste energy. **Figure 12.9** shows the energy available in combustion products for natural gas at different excess air levels as the temperature drops from 600°F to the point of flue gas condensation and full recovery of latent and sensible heat.

For example 400°F flue gasses from a natural gas fired boiler contains 18% of the total energy from the HHV of the input fuel in the form of both sensible and latent heat. Almost 62% of this energy which would otherwise be lost up the stack, requires

cooling the exhaust gasses below the ADP and water dew point. This temperature is a function of excess air (**Figure 12.10**). For natural gas with 10% excess air the initial condensation temperature is 137°F.

The condensation of flue gasses is also a function of temperature as well as excess air as shown in **Figure 12.10**. Note that after the initial condensation temperature is achieved, the percent of water condensed depends on lowering the temperature even further, ranging roughly from 140°F to below 80°F.

Boiler efficiencies above 95% are possible if the moisture in flue gasses can be condensed, **Figure 12.11** shows how the recoverable energy increases dramatically with the condensation of flue gasses. The straight line characteristic of sensible heat recovery bulges from a 2-5% efficiency increase to 11-15% with latent heat recovery.

Various fuels, because of their hydrogen content, offer different opportunities for waste heat recovery as illustrated in **Figure 12.12**. It shows at 450°F 18% for natural gas, 15% for coal-water slurry, 14% for no.2 fuel oil and 11% for coal.

Indirect-Contact Condensing Heat Exchanger

The indirect-contact condensing heat exchanger is generally fabricated from corrosion-resistant materials like teflon and glass.

Since teflon can only be extruded over smooth surfaces and glass tubes cannot be fabricated as finned tubes, an indirect

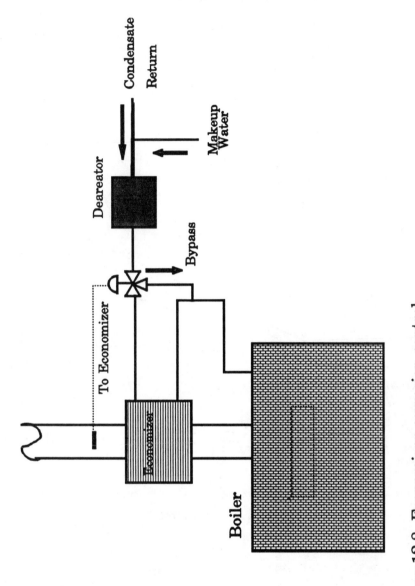

Figure 12.8, Economizer corrosion control by feedwater bypass.

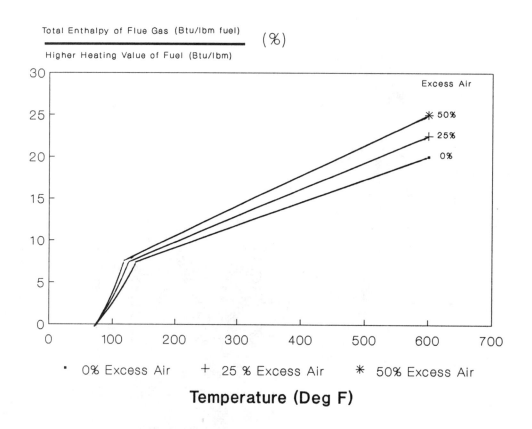

Figure 12.9. Enthalpy of combustion products as a percent of the higher heating value of natural gas as a function of temperature and excess air.

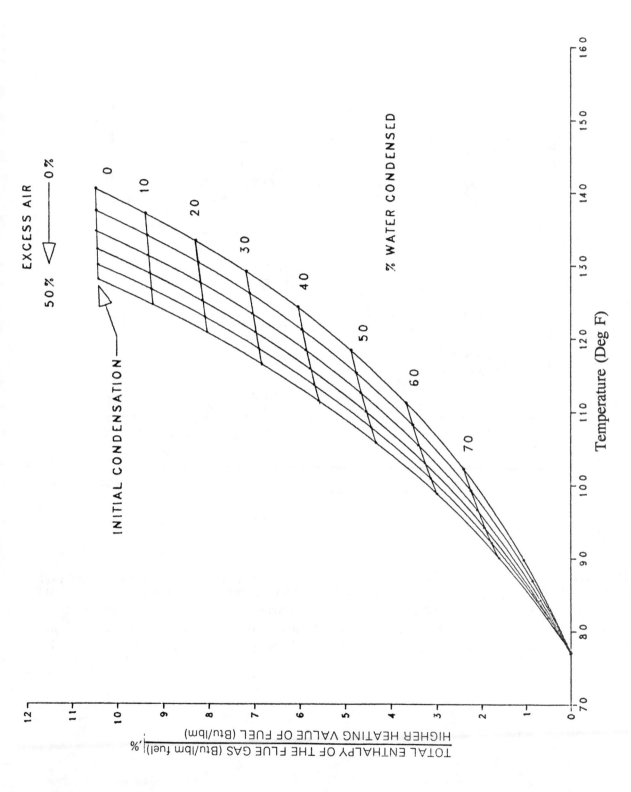

Figure 12.10 Enthalpy of combustion products during condensation, as a function of temperature and excess air for natural gas.

233

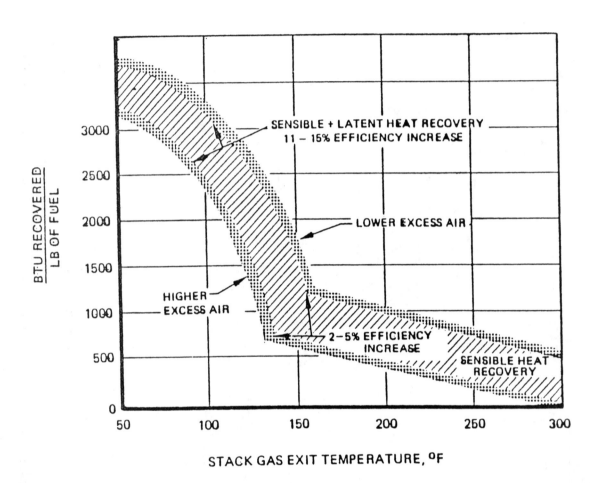

Figure 12.11, This figure demonstrates the heat recovery potential from a natural gas fired boiler by reducing exit gas temperature below the condensing point where latent heat is given up by the flue gas. The potential to recover heat depends on the hydrogen ratio of the fuel and excess air. Typically for natural gas boilers, a 11 to 15 percent efficiency increase is possible.

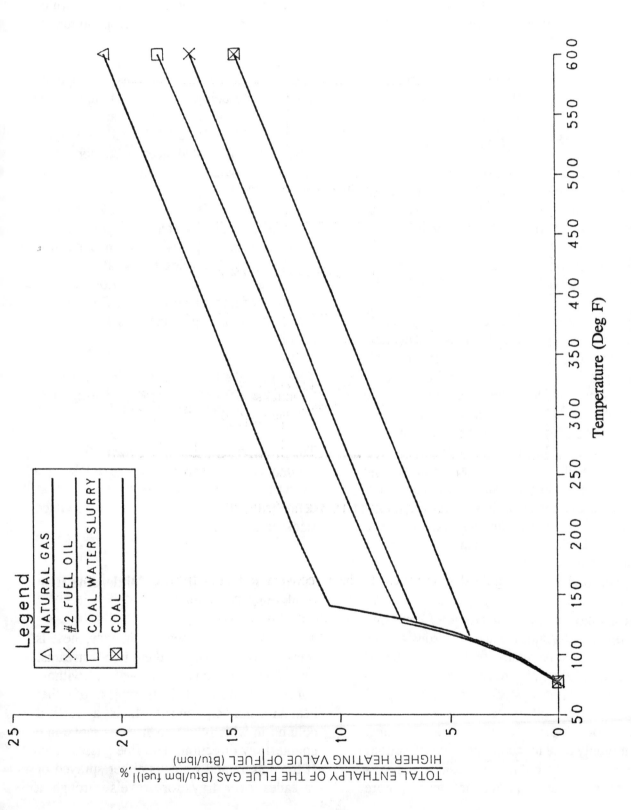

Figure 12.12, Flue gas enthalpy as a function of temperature for various fuels.

contact heat exchanger requires a greater number of tubes and will occupy a greater volume. However, the weight may not be greater since thinner tube walls can be used.

Glass tube heat exchangers are limited to applications where the flue gas temperatures do not exceed 400°F and the water pressure does not exceed 50 psig.

Teflon can be extruded over tubing as a thin film (.015 inch). It can operate with flue gas temperatures up to 500°F, but 400°F is recommended for continuous operation.

Teflon coated heat exchangers are capable of raising water temperature to 200 to 250°F.

Another less common heat exchanger uses stainless steel tubes. Stainless steel heat exchangers were installed in 25 hospitals with gas fired (very low sulfur fuel) boilers, and many of these heat exchangers have operated trouble free for more than 8 years. This type 304 stainless steel has been proven to be a durable material in the stack gas environment for natural gas boilers. The presence of chlorides can cause metal failure due to stress corrosion however. A common source is cleaning solvent vapors, so storage of chloride containing material near the boiler combustion air inlets could cause problems.

Metals have a very wide range of corrosion resistance. As sulfuric and sulfurous acid is the most likely attack to be encountered, material selection must take this into account. Stainless steel, for example does not stand up well under this attack. Carbon steel or "open hearth" steel is most commonly used for economizer construction, giving long and reliable service. It has a corrosion rate superior to many more expensive alloys. Cor-Ten may be another good choice for economizer material, which has a corrosion rate corresponding to titanium.

The most prevalent systems to operate continuously below the ADP to capture the large amount of latent heat usually lost with other systems include borasilicate glass tubes, ceramic coated steel or copper tubes, and teflon-coated copper tubes.

Since corrosion is not a factor with these materials, stack gas can be cooled to well below the traditionally recommended safe temperatures. Lowering temperatures below the water vapor dew point promotes the recovery of latent heat from the flue gas and the recovery of large quantities of low grade energy.

The higher the hydrogen ratio in the fuel the higher the efficiency of the condensing heat recovery unit (**Table 1.1**).

The indirect-contact condensing heat exchanger also acts as a stack gas "scrubber". Researchers report that condensing heat exchangers greatly reduce stack emissions.

Direct Contact Flue Gas Condensing Heat Exchanger

In a direct contact heat exchanger, heat is transferred between the two streams, typically flue gas and water, without intervening walls. This is typical of other heat transfer equipment. It is a vertical column in which the two streams move in a counterflow direction. The flue gasses enter at the bottom and water is either sprayed or cascades over trays or travels through a packed bed from the top to the reservoir at the bottom of the column.

The direct contact heat exchanger is very efficient because there are no heat exchange surfaces exposed to clogging, corrosion or fouling. The elimination of an interfering wall greatly increases the heat transfer rate.

The direct contact heat exchanger is an ideal candidate for transferring latent heat from flue gas because the spraying of fine "mist-like" water droplets can produce a large heat transfer surface in the presence of relatively small temperature differences between heating and cooling streams.

The packed tower design also enhances the available heat exchange surface, providing a highly effective scrubbing action which improves heat transfer efficiency and raises reservoir temperatures.

Figure 12.13 shows a direct contact flue gas condensing heat recovery system in operation. The primary water circuit includes the piping from the reservoir, a pump and the spray nozzles at the top of the column. A heat exchanger is used to transfer heat from the primary system to a secondary system which could be any low temperature water use such as domestic hot water, boiler make up feed water or heating applications.

Figure 12.14 shows the three types of columns used for direct contact heat recovery. The open spray tower presents the least obstruction to exhaust gas flow, less than one tenth inch of water column, but has the lowest reservoir temperature from 100°F to 110°F.

The tray type tower has a resistance to gas flow between 0.5 and 1.0 inches of water column and an outlet temperature from 130°F to 140°F.

The packed tower type has excellent heat transfer but also has a high pressure drop in excess of 5 to 10 inches water column, but the reservoir temperature is as high as 150°F depending on conditions.

Heat Pipe

A heat pipe is composed of a sealed pipe partially charged with water, freon, ammonia or other suitable substance. It can be divided into three sections (**Figure 12.15**), the evaporator section, the adiabatic section and the condenser section.

In the evaporator section, heat is absorbed by the internal working fluid by evaporation. Pressure differential causes the vapor to flow to the cooler condensing section where the latent heat is given up to the cooler environment condensing the working fluid.

The internal circumference of the heat pipe is lined with a thin layer of wicking or mesh type material and the working fluid migrates by capillary action back to the evaporator section where the cycle is repeated.

A thermosyphon type heat pipe does not utilize a wicking material, instead the pipe is placed in a vertical position with the condenser placed above the evaporator section. As the vapor condenses, it returns to the evaporator section by gravity. Even with wicking material it is advantageous to locate the evaporator section below the condenser so condensed fluid is aided by gravity.

Heat pipes have the advantage that they can be constructed of corrosion resistant

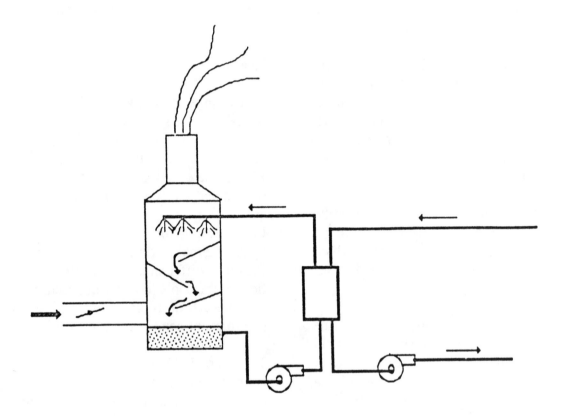

Figure 12.13, Direct contact flue gas condensing heat recovery tower with heat exchanger.

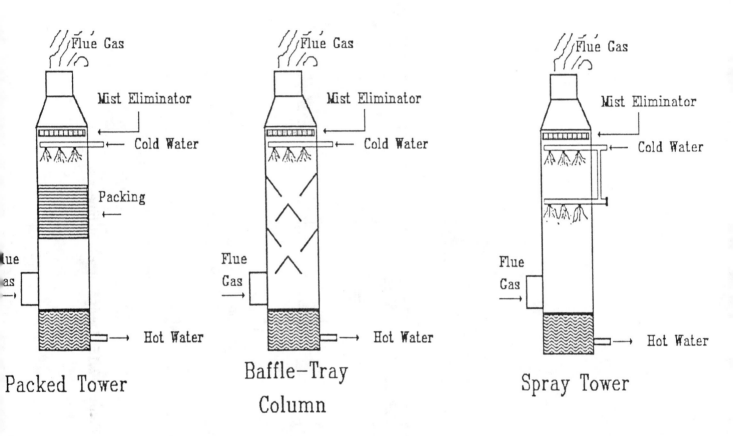

Figure 12.14, Three popular designs of direct contact flue gas condensing heat recovery tower.

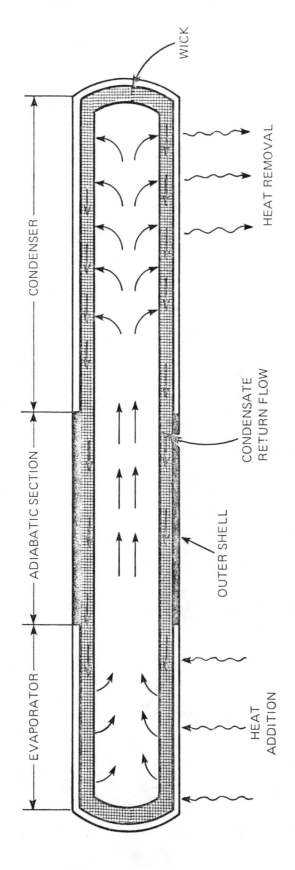

Figure 12.15, Design of a heat pipe.

materials. They are mostly applied to transfer heat from one gas stream to another, but have been used successfully in fluid streams also. The gas streams must be adjacent to each other because of the loss of performance from routing streams of working fluid and vapor through an unusually long adiabatic section. Heat pipes can be compact, operate at low temperature differentials and have no moving parts to malfunction.

Plate Type Heat Exchanger

Plate type gas-to-gas heat exchangers can be used in counterflow or crossflow through adjacent passages separated by heat conducting walls (**Figure 12.16**). It is possible to preheat combustion air, process fluids, building make up air, etc. by this type of heat exchanger. It is a relatively simple system, but the gas streams must pass in close proximity.

One manufacturer has overcome several problems with this type of exchanger. A unit has been designed to be torqued together using resilient mountings, eliminating thermal stress and cracked welds associated with expansion and thermal shocks. Also, these plate units can be constructed of corrosion resistent materials or coated with teflon to insure a long useful life. These units are also designed to benefit from latent heat recovery and to cope with acid corrosion and condensation problems.

The Heat Wheel

The heat wheel is a regenerative type of heat exchanger which extracts heat from one source, briefly stores it and then releases it to a cooler stream (**Figure 12.17**). It consists of a large rotating wheel frame, packed with a heat absorbing matrix. Its primary application is for combustion air preheating. As it rotates it passes through the hot section where the matrix or plates are heated by the flue gasses, then on the other side of a sealing section it rotates through a cold air section giving up its heat.

Heat wheels can exceed 50 feet in diameter. Matrix materials can include aluminum, stainless steel, and ceramics for higher temperatures.

Disadvantages include cross contamination of the two streams due to sealing and purging problems, clogging of passages, gas and air flow restriction and drive motor horsepower. The large size of the units and supporting ducting is another disadvantage. These units can be used to preheat combustion air, but cannot be used to heat building air because of cross contamination of combustion products.

Tubular Heat Exchangers

Tubular heat exchangers are also used for combustion air preheating. The heater is arranged for vertical gas flow through a nest of tubes. The air passes horizontally across the tubes in sections of the air heater defined by built in baffles.

In contrast to regenerative designs, tubular or recuperative air heaters have more severe cold end corrosion problems.

To avoid corrosion, it is necessary to limit the air temperature entering the heat exchanger to a minimum value. This can be achieved by:

　　a. recirculating some of the air from the preheater outlet back to the inlet.

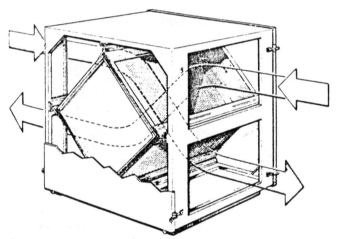

Plate heat exchanger for gases

Figure 12.16, Gas-gas plate heat exchanger.

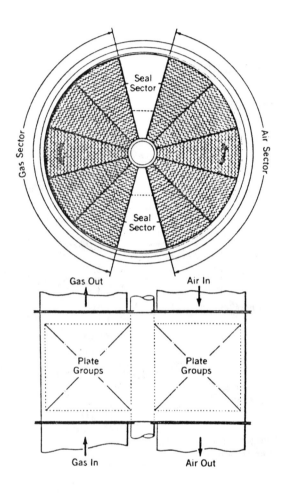

Figure 12.17, Outline of rotary air heater.

Waste Heat Recovery

b. bypassing part of the inlet air around the heating surfaces.

c. using a recuperating steam coil in the air duct upstream of the air preheater.

With high sulfur fuel, the expected life of an air preheater is short. The strategy is to extract as much energy as possible from the flue gas regardless of the corrosion potential and to specify material to keep the problem to a minimum (Table 12.4).

Fuel Type	Incoming Air Temp Range	Material Specification
Oil	190 - 205°F	Carbon steel components.
		Corrosion resistant low alloy steel for the cold-end element.
Oil	155 - 190°F	Corrosion resistant low alloy steel intermediate element.
		Enameled cold end-element.
		Low alloy steel for rotor and structural parts in the cold-end to the same level as the enameled elements.
Oil	Below 175°F	Enameled intermediate element.
		Enameled Intermediate element
		Corrosion resistant low alloy steel rotor and supports to same level as enameled elements.
	Below 155°F	
Bituminous Coal		Carbon steel components.

Corrosion resistant low alloy steel cold end element.

Below 150°F

Pulverized Anthracite	Carbon steel components

Corrosion resistant low alloy steel cold end element.

Gas Fuel Sulfur Free

Table 12.4, Cold-end air preheater temperature material selection guide.

Combustion Air Preheating

Although preheating of combustion air is not generally recommended for smaller boilers, it could be the only way to significantly improve boiler efficiency. As a retrofit, the cost may be high because of bulky ductwork, and possibly a long distance between the stack and air inlet.

Increasing combustion-air temperature worsens emissions of NOx, typically from 20 to 100 ppm for a 100°F rise. Reductions in SO_x, carbon monoxide and particulate emissions result from improved combustion.

In addition to making use of waste heat from flue gasses, lower excess air can be maintained owing to improved combustion at the high inlet air temperature. Higher combustion temperatures also increases heat transfer and reduces sooting.

Gas-to-air heat exchangers are very inefficient in comparison to gas-to-water exchangers. This is due to the superior thermal characteristics of water in comparison to air.

In summary, unless corrosion resistant materials are used, combustion air entering the preheater must be suitably high to avoid cold-end corrosion. A steam coil heater, on the cold side, to raise the temperature of the incoming fresh air is recommended.

Run-around Coil

The typical run-around coil system is composed of two heat exchangers coupled together by the circulation of an intermediate fluid(**Figure 12.18**). The circulating fluid is heated by the hot stream and then piped to the second heat exchanger where its heat is given up to a cold stream.

This system can be applied to transfer heat to combustion air, process air, or building air. Since the heat exchanger requires some temperature differential to transfer heat to or from the intermediate fluid, it is inherently less efficient than a direct exchange between two primary fluids. However this system is relatively simple and more compact than a direct air/fuel gas system. A run-around system eliminates the problem of the close proximity of exhaust and inlet ducts. It is able to transfer heat from one location to another without great retrofit costs.

Organic Rankine Cycle

The Organic Rankine Cycle (ORC) is one method to convert thermal waste streams to electrical energy (**Figure 12.19**). It is a closed loop system filled with an organic fluid having a low boiling temperature. The working fluid is vaporized at an elevated pressure by the waste heat stream in a boiler.

The vapor expands across a turbine which is directly coupled to an electrical generator. The low pressure exhaust vapor from the turbine is condensed and pumped back to pick up more waste heat.

Both initial investment and annual maintenance can be relatively high. The application of an ORC system would depend heavily on the importance and value of electricity as a commodity.

Heat Pump

Heat pumps have the ability to raise low-temperature energy to a higher temperature level. The most common heat pump system is a closed loop filled with refrigerant (**Figure 12.20**).

The heat pump cycle begins with a compressor raising the temperature of the working fluid, in vapor form. This hot vapor goes to a heat exchanger (condenser) where its heat is transferred to heat air or water. The high pressure liquid refrigerant then passes through an expansion valve and its pressure is suddenly lowered causing it to cool. In the evaporator, this low temperature vapor absorbs energy from the waste heat stream raising its temperature. The refrigerant vapor once again enters the compressor and the cycle continues.

The effectiveness of all mechanically driven vapor compression heat pumps is specified in terms of Coefficient of Performance (COP) defined as:

$$COP = \frac{\text{Useful thermal energy output}}{\text{Work input to the compressor}}$$

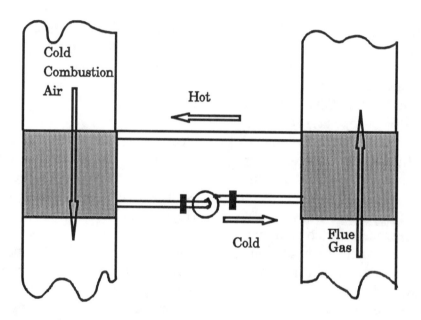

Figure 12.18, Run-around heat recovery system.

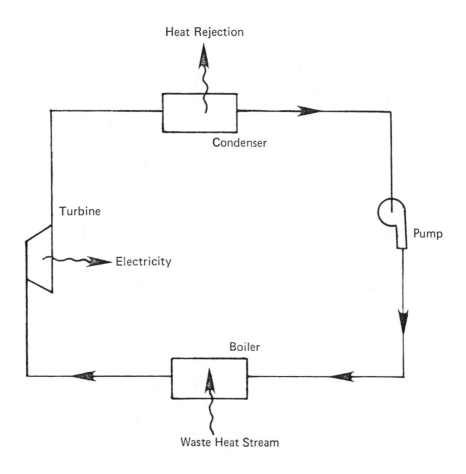

Figure 12.19, Organic Rankine cycle system.

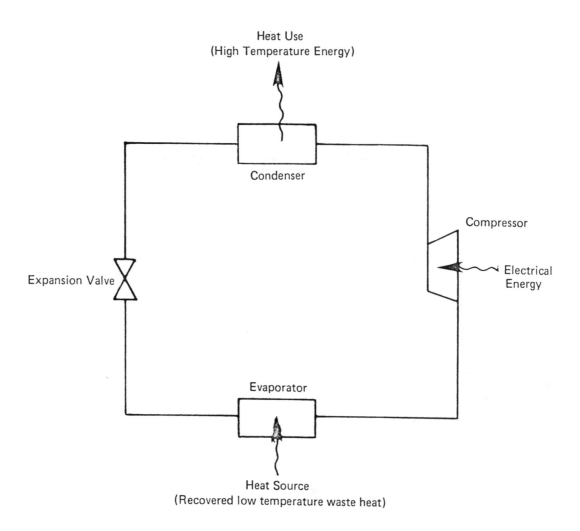

Figure 12.20, Closed-Cycle vapor compression heat pump.

Waste Heat Recovery

A COP of 5 means using one unit of work input (in the form of electrical energy to power a compressor) to deliver 5 units of heat output, of which four units come from the waste heat source. All five output units being at the raised output temperature.

The heat pump is becoming increasingly popular because of this ability to raise the temperature level of recovered energy.

Although heat pumps are most often driven by electrical motors, heat pumps driven by combustion engines have the advantage of being powered by fuels costing a fraction of electricity and the added advantage of using waste heat generated by the engine.

Both turbines and internal combustion engines can recover 70% to 80% of their own waste heat to add to system efficiency. Driving a compressor with a fuel-fired engine rather than an electric motor becomes attractive when the electricity costs are high and natural gas or fuel oil costs are low.

Preheating Air or Water?

Due to the superior heat transfer qualities of water, preheating feed water is more efficient and economical than preheating combustion air. However, air is always available at ambient temperature with good potential for accepting flue gas heat.

Fire Tube Boiler Air Preheating

In general, fire tube boilers are not designed to operate with preheated combustion air, and it would be impractical to attempt to improve boiler efficiency with this type of system.

Which Option is Best?

Based on a number of general assessments considering all facts involved in economic benefits, engineering criteria, installing and operating waste heat recovery systems; three primary systems have evolved as being very good prospects for waste heat recovery. The waste heat recovery systems which showed the most promise and applicability are:

    a. The conventional economizer
    b. The indirect-contact condensing heat exchanger
    c. Direct contact flue gas condensation heat recovery

Table 12.5, summarizes the merits and limitations of various heat recovery alternatives.

Waste Heat Recovery

Equipment Type	Merits	Limitations
Economizer (Gas-Liquid)	Well developed and understood, easily tailor made with a variety of materials	Cold-end corrosion, water temperature approach to steam temperature 40°F
Waste Heat Boiler (Gas-Liquid /Steam)	Compact, high heat transfer easy to clean, robust, can operate in fouling environment	Not generally internal to process, must link with steam distribution system, most appropriate to high temperature exhaust streams (>600°F)
Direct Contact Flue Gas Condensing (Gas-Liquid)	Good heat transfer, simple, recovers sensible heat and latent heat. Efficiencies over 95% possible.	Lower grade heat recovery 120°F to 150°F Good for natural gas; less appropriate for oil/coal.
Indirect contact Flue gas Condensing Teflon (Gas-Gas) (Gas-Liquid)	Corrosion resistant, reliable under under thermal and mechanical shock, recovers latent heat. High efficiencies possible.	Teflon cannot be finned, operating range must be below 400 - 500°F.
Indirect Contact Flue Gas Condensing Glass, (Gas-Gas) (Gas-Liquid)	Good resistance to corrosion, easy to clean, tailor made, can recover latent heat, no cross contamination, can recover latent heat, high efficiencies	Large area of borasilicate glass needed, temperature limitation (< 400°F),sensitive to mechanical vibrations and shocks.
Plate Exchangers (Gas-Liquid) (Gas-Gas) (Liquid-Liquid)	No cross contamination, easy to install on-line cleaning, compact, temperature differentials can approach 2°F. Condensing units available.	Only small pressure differentials can be tolerated, large, inconvenient ducting. Difficult to clean in some circumstances. Subject to thermal and mechanical shock.
Heat Pipe Heat Exchangers (Gas-Gas)	No moving part , no cross-contamination, large pressures can be tolerated, compact. Various working fluids suite different temperature ranges.	Gas streams must be close.
Run-around Coil (Gas-Gas)	Covers large distances eco-nominally, no cross contamination, easy to design and install.	Thermal efficiency less than 65%; moving parts, temperature limitations. Freeze protection may be needed. Electrical load controls needed.
Rotating Preheater Heat Wheel (Gas-Gas)	High operating efficiency 85%, low pressure drop, can recover latent heat. Very large sizes. Operating temperature to 800°F.	Cross contamination possible, moving parts, wear, large ductwork. Low differential pressure tolerance between gas streams can be a challenge.

Table 12.5, Merits and limitations of heat recovery equipment and approaches.

Chapter 13

Steam System Optimization

Steam System Optimization, a huge opportunity for savings

According to the Department of Energy, nearly 160 billion dollars is spent creating steam in a single year. Steam generation accounts for fully one half of the industrial and commercial energy dollar. When you consider that steam plays a major role in generating electricity this number can be much higher. Energy conservation measures described in this chapter can play a significant role in saving over 15 billion dollars annually which is lost in steam distribution systems.

The cost of operating a steam system includes:
  a. Boiler operating costs
  b. Steam distribution system losses
  c. Trapping system losses
  d. Condensate return system losses
  e. Operation and maintenance of the systems involved.

Boiler Plant Optimization

Boiler plant operating costs and losses have been covered in the earlier chapters of this book. An important fact to keep in mind is that every Btu wasted in the steam distribution system has to be replaced by the boiler and the boiler has large losses of its own. If a boiler is 80% efficient, then at least 20% of the original fuel used to generate steam for the distribution system is lost and does no useful work.

Getting the Most Energy from Steam Systems

Energy in steam consists of sensible and latent heat (**Figure 13.1**). It is apparent that a significant amount of heat is invested in bringing water up to boiling temperature. This sensible energy may be lost through a phenomena known as "flashing" when hot condensate is formed at a high pressure and escapes to a lower pressure through the steam trapping system.

Figure 13.2 shows what happens to the condensate when it escapes from a higher pressure to a lower pressure. Illustrated here is a trap for a heat exchanger which is forming condensate at the rate of 100 pounds per hour. The system pressure is 100 psi and the corresponding steam temperature is 353°F. This 100 Lb of condensate contains 32,500 Btu.

On the low pressure side of the trap, which is atmospheric pressure in this case, the saturation temperature is 212°F. Of the 100 pounds of condensate, 14.8 lb will form flash steam with a heat value of 17,100 Btu. The remaining 85.2 pounds of condensate formed will contain 180 Btu per pound for a total heat content of 15,400 Btu. Notice that the 32,500 Btu in the condensate on the high pressure side of the trap has formed steam and water with the same total heat content of 32,500 Btu.

BTUs in a Pound of Steam

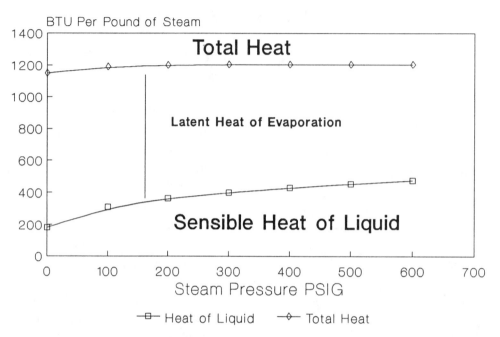

Figure 13.1, Heat in one pound of saturated steam at various pressures.

FLASH STEAM TABLE

INITIAL PRESSURE PSIG	TEMP OF LIQUID	% FLASH AT ATMOSPHERIC PRESSURE	PERCENT OF STEAM FLASHED AT REDUCED PRESSURE				
			5PSI	10 PSI	15 PSI	20 PSI	25 PSI
100	338	13.0	11.5	10.3	9.3	8.4	7.6
125	353	14.5	13.3	11.8	10.9	10.0	9.2
150	366	16.0	14.6	13.2	12.3	11.4	10.6
175	377	17.0	15.8	14.4	13.4	12.5	11.6
200	388	18.0	16.9	15.5	14.6	13.7	12.9
225	397	19.0	17.8	16.5	15.5	14.7	13.9
250	406	20.0	18.8	17.4	16.5	15.6	14.9
300	421	21.5	20.3	19.0	18.0	17.2	16.5
350	435	23.0	21.8	20.5	19.5	18.7	18.0
400	448	24.0	23.0	21.8	21.0	20.0	19.3
450	459	25.0	24.3	23.0	22.0	21.3	20.0
500	470	26.5	25.4	24.1	23.2	22.4	21.7
550	480	27.5	26.5	25.2	24.3	23.5	22.8
600	488	28.0	27.3	26.0	25.0	24.3	23.6
BTU PER POUND OF FLASH STEAM		1150	1155	1160	1164	1167	1169
DEG F OF FLASH STEAM & LIQUID		212	225	240	250	259	267
STEAM VOLUME CUFT/LB		26.8	21.0	16.3	13.7	11.9	10.5

Table 13.1, Percent flash steam generated by condensate or boiler water when pressure is lowered.

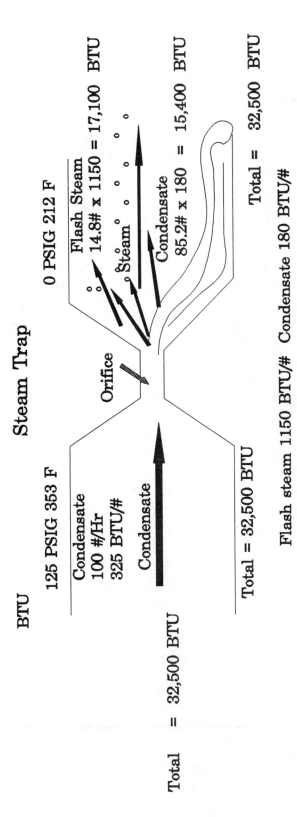

Figure 13.2, Flash Steam Energy Balance.

Steam System Optimization

Table 13.1 shows the percentage of steam that will form at the new lower pressure. This low pressure flash steam will, in fact, contain more heat than the condensate. There is a great potential for half of the energy going through the trapping system to be lost as flash steam, especially in condensate recovery systems that vent to atmosphere.

The higher the steam system pressure, the higher this flash steam loss will be, so one important factor that should be examined is how to reduce this loss.

Table 13.2 shows that the higher the steam system pressure, the less latent heat is available in the steam. This latent heat is what does the work, the rest is lost through the trap. If we were able to drop the working pressure on a piece of equipment from 125 psig to 5 psig, the latent heat would be 83.5% of the total energy instead of 73.5%.

Pressure psig	Latent Heat BTU/Lb	Percent of Total Heat
125	868	73.5
50	912	77.5
5	961	83.5

Table 13.2 shows why steam pressure should be reduced at the point of use. It conserves the loss of energy through the steam trap.

Significant steam savings can be realized by reducing steam pressure as low as possible using pressure reducing valves in new as well as existing installations.

As shown in **Table 13.3**, the percent heat savings available from reducing pressure is significant. For example, there will be a 13% savings by reducing the pressure from 150 to 20 psig.

Reduced Pressure	Original Pressure (PSIG)		
	125	150	200
	(Steam Savings %)		
100 psig	2.1	3.6	6.8
75 psig	4.4	5.9	9.0
50 psig	7.2	8.6	11.6
20 psig	11.5	13.0	15.8
10 psig	13.7	15.0	17.8

Table 13.3, Heat savings available from reducing pressure at the point of use. (To calculate these numbers, flash steam reduction and the latent heat of the steam and the heat in the condensate at the old and new pressure is considered)

Steam System Optimization

A Rule of Thumb for Optimizing Steam System Pressure:

    a. System steam pressure should be distributed at high enough pressure as is practical to overcome line losses and satisfy the highest pressure user.

    b. Steam pressure should be reduced at the point of use to as low a pressure as is practical.

Two general points to consider in the steam distribution system is, design requirements limit steam velocities to less than 6,000 to 12,000 feet per minute and the system pressure drop should not exceed 20 percent of the total maximum pressure of the boiler.

Heat Transfer Efficiency

Accumulations of air and noncondensible gases in the steam system can also limit steam flow, steam temperature and heat energy release.

Air is present in the system on start up and is also introduced by vacuum breakers on heat exchangers and process equipment. Noncondensible gases are liberated in the boiler by bicarbonates which forms CO_2. Oxygen is carried through the system. These noncondensible gases, when released, flow with the steam and can create heat transfer problems.

The gases cause a temperature reduction by contributing to total system pressure.

Dalton's Law of Partial Pressure states that a mixture of steam and other gases is equal to the sum of the partial pressures. This effectively reduces steam pressure, temperature and energy transfer.

Table 13.4 shows the effect of air on the temperature of a steam-air mixture.

Mixture Pressure	Temperature °F			
	Pure Steam	5% Air	10% Air	15% Air
2	219	216	213	210
5	227	225	222	219
10	239	237	233	230
20	259	256	252	249

Table 13.4, Air-steam temperatures.

Insulating Barriers

There is a second phenomena involved with noncondensible gases in the steam. When steam condenses on the heat exchange surface, the noncondensible gases also accumulate on the surface forming an insulation barrier.

The first one percent of air barrier has the most effect on reducing heat transfer **(Figure 13.3)**. In this figure T_s is the steam temperature and T_w is the water temperature of the heat exchanger.

Steam System Optimization

Heat Transfer Coefficient

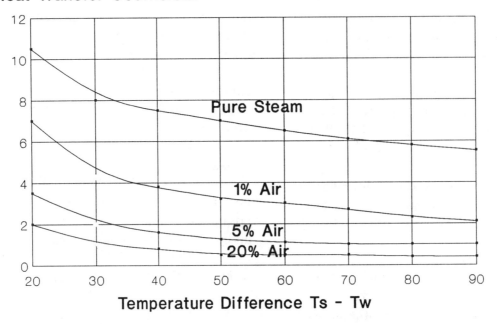

Figure 13.3, Effect of air in steam on heat transfer.

Heat Loss $/Year
4 Inch Gate Valve
(Uninsulated)

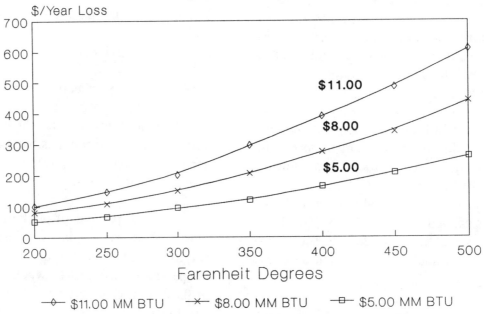

Table 13.4, Annual dollar energy loss cost from an uninsulated 4 inch valve.

Steam System Optimization

These gases take up volume and don't condense into a liquid as readily as steam, hence the term noncondensible. If allowed to accumulate for long periods, they take up enough volume to effectively block steam flow and energy transfer.

Bellows type thermostatic steam traps can be used as automatic air vents on heat exchange equipment. Air and noncondensibles in the system tend to be lighter than the steam and accumulate in quiet zones. If installed at these locations, the thermostatic device can sense the temperature reduction caused by the air.

Batch process cookers, large shell and tube heat exchangers and large steam coils should incorporate automatic air vents to eliminate air accumulations.

Steam Distribution System Losses

Insulation

Steam is distributed through hot pipes which must be kept insulated to prevent excessive loss of heat and for safety. The range of surface temperature can vary from 200 to 500 °F. The bare surface losses can vary from 300 Btu/Hr to 1700 Btu/Hr. The losses from uninsulated surfaces can be impressive. **Figure 13.4** shows the losses from an uninsulated four inch gate valve for one year for steam costing $5.00, $8.00 and $11.00 per million Btus.

Steam users seem to be fully aware of the need to insulate hot surfaces to prevent heat loss. Insulation pays for itself quite quickly, but insulating steam piping means not only the main piping, but also the unions, flanges, valve bodies, steam traps and everything else that is hot.

Removable (lace-up, foam in place and molded etc.), insulating covers are available for valve bodies and other hard to fit shapes. Mechanical steam traps such as float and thermostatic (F&T), inverted bucket (IB) should also be insulated as well as the bodies of thermodynamic disc traps (the covers should be left bare). Only Thermostatic (TS) traps and their cooling legs should be left uninsulated.

Condensate return lines should also be well insulated to insure that all of the heat possible is returned to the boiler plant for additional savings. **Table 13.5** shows heat losses per hour for bare metal temperatures in 80 degree still air.

Loss BTU/SQ Ft/Hr		
Steam Pressure PSIG	Temperature (F)	Heat Loss BTU/SQ Ft Per Hour
15	250	452
35	280	556
50	298	595
75	320	716
100	338	805
125	353	864
150	355	876
175	377	989
200	389	1,051
300	422	1,238
400	448	1,404
500	470	1,551
600	489	1,693

Table 13.5, Uninsulated bare metal heat loss in 80 degree still air for various steam pressures.

Steam System Optimization

Inactive Piping

Inactive or dead piping must be positively isolated from the steam supply, this is often overlooked. Inactive steam piping wastes 100% of the steam supplied as well as being a maintenance burden.

Instead of relying on a hand operated valve, which may not shut tight or be leaking, it is better to blank off the unused part of the system or remove a section of piping to insure positively that this section is not wasting good energy.

External Steam Leaks

Steam leaks to the atmosphere are usually visible, audible or both. Because of the safety hazard they present, they are usually fixed quickly. However many leaks seem to go on being neglected for years. Their large plumes are hard to hide and can be seen for long distances (**Table 13.6**).

Plume Length (Ft)	Cost/Yr @ $7.00 Million Btus
4	$3,300
5	$5,900
6	$10,000
7	$18,000
8	$32,000

Table 13.6, Steam leak energy loss estimates based on steam plume length.

Internal Steam Leaks

Steam plumes are easy to spot. What about the steam leaks you can't see? Leaks inside steam systems are invisible.

Valves in bypasses around steam taps are prime candidates for this type of leakage. They are often opened to assist system start up or to vent excessive air and can easily be forgotten or only closed partially or develop leaks over time.

The need for bypasses should always be questioned. If they are needed for startup it is likely that their real purpose is the discharge of non-condensible gasses faster than a steam trap can handle them. An automatic air vent can do the job better and more economically. A trap could be installed in the bypass line as a back up in case repairs are required for the main trap or to provide more capacity during high loads. This way the liability of internal steam loss can be prevented.

Condensate Return System Losses

Condensate is certainly not waste water. It is water that has been preheated, softened, made suitable for use as boiler feed water, and then distilled.

If it can be returned to the boiler feed system, it can be recycled with little further treatment. By reusing the condensate, you avoid the need for more raw water to be purchased and treated and you can avoid the sewage costs which can be incurred when condensate is allowed to run to the drain. Also, this condensate contains valuable heat energy that was paid for in the boiler fuel; energy that an efficient condensate recovery system will conserve.

Steam System Optimization

The efficiency of a condensate return system can be improved, sometimes dramatically. For example, **Table 13.7** shows annual dollar savings from improving condensate return. Savings are based on steam generation cost of $6.00 per 1,000 lbs. Make up improves from 100% to just 10%, a 90% reduction. Annual dollar savings for three boiler sizes listed include water cost, chemical treatment and energy savings.

100 BHP	250 BHP	500 BHP
$21,900	$54,750	$109,500

Table 13.7, Annual cost reduction from improving condensate return.

The Basic Challenge

The condensate system is perhaps the most ignored and least understood system in a steam using facility. An examination of the dollar savings listed in **Table 13.7** suggests that it might be profitable to spend some time studying your condensate systems.

One of the most useful instruments for this job is a makeup water meter. It can be used to detect system leaks, malfunctioning system components as well provide an excellent basis for calculating wasted energy.

The rule of thumb on condensate return energy is:

For each 11°F the water temperature is cooled, it loses 1% of efficiency for the system.

The challenge is to get this heat back to the boiler plant. The flash steam formed as the condensate leaves a trap is of first concern, approximately half the energy in the trap exhaust is the energy in this flash steam. In an open system it can escape to atmosphere through a vent in the receiver tank.

If the condensate piping is not adequately insulated, additional losses will occur.

Some pumping systems cannot handle condensate above 180°F because of cavitation. There may be other operational problems that have lead to lower condensate return temperatures. Some plants have actually devised ways to cool condensate so they could get their pumps would work.

Good management of the condensate systems has additional benefits:

    a. Reduction in replacement water costs.

    b. Reduction in water treatment chemicals.

    c. Reduction in fuel consumption used in preheating the make-up water.

One solution to getting the hot or even boiling condensate back to the boiler room is the steam pressure powered pump (**Figure 13.5**). It collects condensate and discharges it when the liquid level reaches a certain point. It uses system steam pressure for the pumping action, a small investment for high overall energy savings.

The condensate pressure powered pump eliminates many problems connected with handling hot condensate with electric pumps. When the condensate goes above 190°F they tend to cavitate, forming steam in the suction end of the pump. One common solution had been to cool the condensate to a temperature where the old pumps could handle it. Another common practice is to dump the condensate down a convenient drain.

This type of pump is reported to use only 3 pounds of steam for every 1,000 pounds of liquid pumped. When exhaust is vented back in a closed system, the steam is recovered and the cost of operation is negligible.

Acid Corrosion and Oxygen Attack in Piping Systems

A great part of condensate recovery system failure can be traced to a feed water treatment problem that allows carbon dioxide to enter the distribution system piping. Oxygen also causes problems but most are confined to the boiler and economizer.

Carbon dioxide that carries over in the steam forms carbonic acid which has the capacity to combine with one and a quarter pounds of steel per pound of CO_2, forming a groove in the bottom of the piping. Over years this can eat up a lot of metal and cause countless problems. Oxygen pitting and scale formation can also destroy piping and boiler tubes as well as interfere with heat transfer and the operation of pressure reducing valves and trap mechanisms.

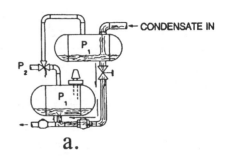

a.

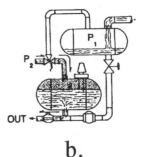

b.

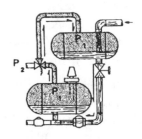

c.

Figure 13.5, The operation of a pressure powered condensate pump: (a) is the fill cycle, (b) is the pumping cycle and (c) is the venting cycle.

Chapter 14

Steam Traps

When steam expends its energy, by giving up its heat or by doing work, it condenses back into water **(Figure 14.1)**. This water must be removed from the pathway of the steam so it will not interfere with the function of the steam system. Once removed, the condensate, pure hot water, should be returned to the boiler where it is again heated to produce more steam.

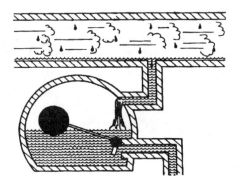

Figure 14.2, Float trap is trapping steam while allowing condensate to discharge.

The BTU's which are released by steam in heating, and process applications and by pipe radiation loss causes the steam to condense and form droplets of water that can quickly combine into larger masses. If this condensate is not effectively removed (trapped), it can reduce the efficiency of heat transfer equipment by a phenomena known as waterlogging.

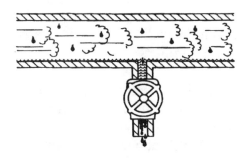

Figure 14.1, Condensate formation in steam distribution line. A valve is shown draining line instead of a steam trap.

The Purpose of Steam Traps

The job of removing condensate is handled by the steam trap **(Figure 14.2)**. The steam trap's job is to remove condensate while preventing steam from escaping from the distribution system. It must discharge this condensate from a higher to a lower pressure. To do this job, it is designed to differentiate steam from condensate, usually by reacting to temperature, density or thermodynamic properties.

Condensate accumulating at the bottom of a pipe is swept along by the high velocity steam **(Figure 14.3)**. As the water moves through the pipe collecting additional droplets, forming a larger and larger slug, it develops a high level of energy and can cause serious damage through the phenomena known as "Waterhammer". Severe waterhammer can burst the wall of a pipe, possibly causing personal injuries and damaging other equipment in the area, it is usually accompanied by a sharp metallic noise.

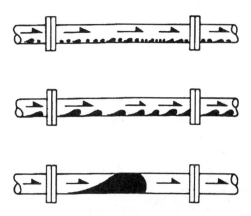

Figure 14.3, Condensate accumulates in distribution piping and forms a slug which is carried along by the steam at high velocity.

Severe damage can occur when the incompressible slug of water is forced to change direction by a pipe bend, fitting or valve. It produces a shock wave of tremendous momentary force which can cause great physical damage. It is a function of the mass of water and the velocity squared, so the greater part of its energy comes from the high velocities (6,000 - 12,000 feet per minute) achieved in the steam piping.

Some steam boilers have a tendency to produce low quality steam, which contains an appreciable amount of moisture and possibly boiler chemicals and contaminated boiler water. This introduces water into the boiler header and distribution piping. Depending on the severity, this can overload the distribution and trapping systems.

To detect this problem a Throttling Calorimeter can be used to test for steam quality (percent moisture in the steam). This instrument measures the temperature of a very small amount of steam discharging from the steam main. Dry steam will become superheated due to the pressure drop. If there is any moisture present, the temperature of the discharge will be reduced. Measuring temperature and comparing it to the maximum possible temperature will indicate steam quality.

Other methods that can be used are:

    a. Ion exchange

    b. Conductivity

    c. Sodium tracer flame photometry

    d. Specific ion electrodes.

These methods determine the solids content of the steam, including the solids carried over by water droplets.

The throttling steam calorimeter is the only direct way to determine steam quality and is most accurate below 600 psi and where moisture content is above 0.5 percent.

When low water quality is detected, it indicates a problem with the water level control, feedwater treatment, blowdown cycle, boiler sizing or steam drum internal problems. Boiler and steam generator operating procedures may also have to be investigated and changed.

All of the steam generated by the boiler must eventually leave the system as condensate, except for leaks and steam directly used for some purpose. It is obvious that if the steam can somehow escape from the system without going through a trap as condensate, the system will be inefficient to that degree. Properly working traps insure that steam gives up its energy efficiently.

gases, upon release, flow with the system and can create energy problems. When steam condenses, these gases migrate to the heat exchange surface, forming a insulating film **(Figure 14.5)**. This film can be a very effective insulator, usually only a very small percentage is needed to cause a big problem with heat transfer.

The gases also take up volume and don't condense into liquid as readily as steam, that's why they are called noncondensibles. If allowed to accumulate for long periods, they can take up enough volume to effectively block steam flow and almost all energy transfer. When condensate drainage is blocked, dangerous water hammer can occur.

Trapped air can cause heat transfer problems, especially in equipment, such as shell and tube heat exchangers, where there is room for it to accumulate at high points or in areas of low velocities. This trapped air will cause a cold spot, that's how you can find them. This would be a good indication that an air vent is needed.

When allowed to cool in the presence of condensate, carbon dioxide can combine with water to form carbonic acid. Since gas accumulation causes a temperature drop, acid formation is highly probable in any standing condensate.

The corrosion of iron forms a soluble bicarbonate which leaves no protective coating on the metal. If oxygen is also present, rust forms and CO_2 is released, which is now free to cause more corrosion.

Once gases become dissolved, they should be removed, but while in the system they usually contain some damaging acids. The

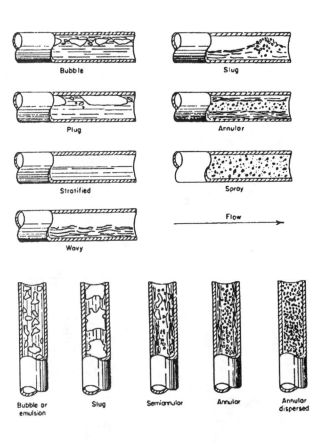

Figure 14.4 Shows several forms of steam and condensate mixing in piping.

Air and Noncondensibles in the Steam System

Accumulations of air and noncondensible gases in the steam system can limit steam flow, steam temperature and heat energy release. Air is present in the system on startup and is introduced through vacuum breakers.

Noncondensible gases are liberated in the boiler. Carbon dioxide and oxygen are dissolved in boiler feedwater as carbonates and bicarbonates. These noncondensible

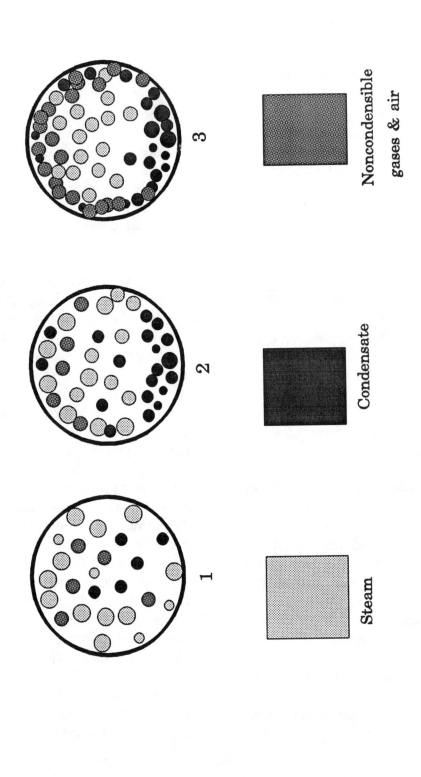

Figure 14.5. Reduction of heat transfer by noncondensible gases: (1) a steam line containing steam, condensate and gases, (2) noncondensibles are migrating to the heat exchange surface as steam condenses and (3) an insulating film is forming (darker shade) at the heat transfer surface and condensate is forming in the bottom of the line.

corrosion they produce causes piping leaks, damaged traps and valves and formation of insulating scales.

System Protection

One application, of many that are needed in a steam system, is the removal of condensate from steam strainers **(Figure 14.6)**. Strainers are used to protect pressure reducing valves and traps and are generally installed to keep the system clean. The strainer body is a low point in the system which accumulates condensate naturally, reducing the effective area of strainer screen. In the case of a valve down stream that has been closed, some condensate will be picked up by the flow when it is opened, impacting the valve seat with dirty condensate.

Installing an inverted bucket steam trap on the strainer blowdown will keep the condensate drained. This will free strainer surface from blockage improving strainer performance. This may be the reason why some valves have perpetual problems.

How Steam Traps Fail

Steam traps are subject to harsh operating conditions and like all mechanical devices, their moving parts are subject to wear, corrosion and eventual failure.

Most traps function intermittently. During the closed portion of their cycle, the leg of water that accumulates encourages the formation of a mildly corrosive acid and the cycling itself causes temperature fluctuations which accelerate the problem.

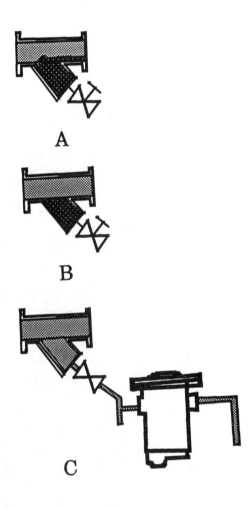

A

B

C

Figure 14.6, installing a steam trap on a strainer to prevent valve damage.

A typical trap may open and close several million times a year and some wear, malfunction and outright failure is inevitable. Steam losses conservatively exceed 15 billion dollars annually.

Steam trap leaks are a form of invisible steam leak. Rather than arrive at the point of use, the steam escapes to the condensate return system without accomplishing useful work.

Steam Traps

There are four essential causes for steam trap leaks: **(Figure 14.7)**

    a. The trap responds too slowly, not closing fast enough to prevent the escape of some steam on the closing cycle.

    b. The trap leaks in the closed position because of either a defect in the valve closing mechanism or in the sealing surfaces allowing steam to leak through.

    c. The trap fails to close completely, because of mechanical failure.

    d. The trap fails open, allowing steam to blast through the escape orifice.

All steam traps will eventually fail. Most traps fail because mechanical parts wear out through normal operation. Others will fail because of the flashing of condensate as it passes through the trap, which can wire draw the valve seat. Still others will fail because of the stresses of steam service such as water hammer, rapid temperature fluctuations, carbonic acid corrosion or general fatigue of operating components due to millions of operational cycles each year.

Steam traps are small, relatively inexpensive and short-lived components of the steam system which represents a very large opportunity for savings in almost any plant.

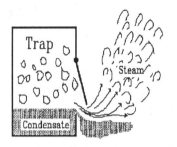

a.

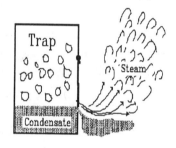

b.

c.

Figure 14.7, Various ways a steam trap can waste steam: (a) is either failure to close completely or closing too slowly, (b) is leaking in the closed position and (c) trap fails open.

Steam Trap Types and Characteristics

Effective steam trapping practices have a direct impact upon the efficiency of any steam or condensate system.

To ensure maximum steam trap life and steam system efficiency, steam trap type and size must be properly matched to each application. It is important to emphasize there is no "UNIVERSAL" steam trap which will provide efficient operation for all applications.

Most facilities will require several types of steam traps in various sizes to provide efficient steam trapping. The user must have knowledge of the various types of traps available to properly select and specify the traps to be used.

Categories of Steam Traps

All steam traps are designed to distinguish between condensate and steam. To be effective they must stop or trap the steam and release condensate. The physical differences between steam and condensate has lead to a variety of approaches to steam trap design.

There are three basic categories of steam traps, each using a different principle to differentiate between steam, non-condensible gas and condensate. These are:

Mechanical, which are operated by differences in density between steam and condensate.

Thermodynamic, operated by kinetic energy.

Thermostatic, operated by temperature.

Mechanical Steam Traps

Mechanical steam traps are operated by the difference in density between steam and condensate. All use a float in some form to operate the valve to let condensate drain from the system.

Mechanical traps include Inverted Bucket, Open Bucket and Float and Thermostatic (F&T) types. All can operate at or near steam saturation temperature because trap operation is dependent on the density difference between steam and water which is about 700 to 1. **Figure 14.8** shows the density relationship between water and steam.

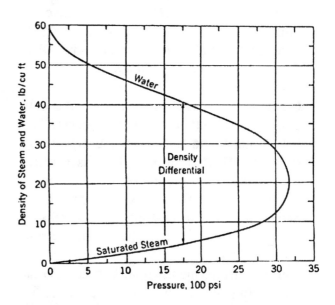

Figure 14.8, Water steam density relationship.

All mechanical traps respond quickly to changing loads and most have large turn down ratios. However, without a separate mechanism to discharge air and non-compressible gasses, mechanical traps have a limited venting capacity.

Since all mechanical traps depend on a condensate level inside the trap for proper operation, they are position sensitive. Some will fail to operate when installed even a little off vertical.

Inverted Bucket Traps

Inverted bucket traps use an open inverted bucket as a float. As with all mechanical traps, inverted bucket traps operate on the difference in density between steam and condensate.

The inverted bucket is usually attached to the valve by a lever mechanism. When the inverted bucket sinks toward the bottom of the trap body, the valve is opened. Since the linkage is designed with some "play", inverted bucket traps can operate in either the cyclic or modulation mode.

Condensate, steam and non-condensible gasses enter under the inverted bucket. The lighter steam and air lift the bucket, closing the valve **(Figure 14.9)**. There is a bleed hole in the top of the bucket to let air and a small amount of steam bleed through so the bucket doesn't get stuck in the closed position.

During startup, the inverted bucket rests on the bottom of the trap body because of its own weight. The valve is open allowing the discharge of cold condensate, air and non-condensibles.

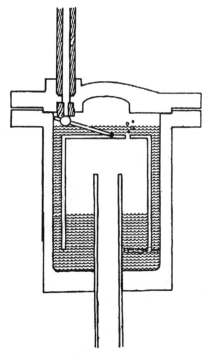

Figure 14.9, Inverted bucket steam trap has trapped steam under the bucket and floated up to close the outlet valve.

As the condensate enters the trap, a liquid level is formed which provides a water seal around the inverted bucket. Since the bucket is filled with air and non condensible gasses, the weight of the bucket is overcome by the buoyant force of these gasses in water, and the bucket rises closing the valve. The air and non-condensible gasses trapped under the bucket are bled through the hole at the top of the bucket, allowing more condensate to enter. The bucket loses buoyancy, sinks and opens the valve.

If the amount of air and non-condensibles entering the trap is greater than the vent capacity of the bleed hole, the trap can air bind and fail closed. Some manufacturers have alleviated this problem by using larger

Steam Traps

bleed holes or thermostatic elements to enhance their venting capacity.

Once the valve is open, discharge continues until all of the condensate has been removed from the system and steam floats the bucket closing the valve (**Figure 14.10**). The trap will remain closed until enough steam has vented through the bleed hole or condensed to allow the bucket to sink, opening the valve and starting the cycle again.

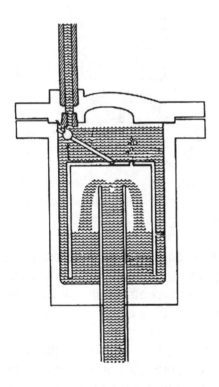

Figure 14.10, Inverted bucket trap with inverted bucket lowered and the valve open and discharging condensate.

The slack or play in the linkage allows the system pressure to snap the valve closed after the bucket has risen within the trap body. System steam pressure acts to keep the valve closed until the bucket has lost enough buoyancy to pull the valve off the

seat. This action is affected by system pressure which requires special consideration given to the sizing of the orifice. Therefore, in order to maintain the proper relationships between the closing force of the pressure on the valve and the opening force of the weight of the bucket, orifice sizes must be designed to match steam system pressure.

Since the orifice must be sized to match the maximum steam system pressure, care must be taken to ensure adequate capacity is still available.

Open Bucket Traps

The open bucket trap is an older and possibly obsolete type of mechanical trap. It is simply a bucket arrangement that pivots around one end when it fills with water, opening a discharge valve. Most open bucket traps have a separate thermostatic air vent at the top.

During start up the bucket rests on the bottom of the trap body and the valve is fully open. The thermostatic element is also open, allowing the discharge of air and gases. When condensate enters the trap the enclosed trap body is filled with condensate causing the bucket to rise shutting off the valve. Condensate continues to enter the trap as the thermostatic element allows air and gases to be dispelled from the top.

The liquid level rises above the open bucket filling it and causing it to sink, opening the valve. Condensate is discharged until steam enters the trap body and the water is forced from the open bucket, making it light enough to float up closing the discharge valve.

Steam Traps

Without the thermostatic element or some other means to clear the trap body of air, the trap will air bind and fail closed.

Float Traps and Float and Thermostatic Traps

Float traps **(Figure 14.11)** are one of the oldest trap designs still in use and are not considered obsolete.

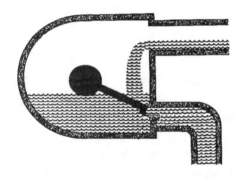

Figure 14.11, Float type steam trap.

A float trap consists of a closed float attached to a valve by mechanical linkage. During start up, the float rests on the bottom of the trap, closing the valve. Air and gases must be removed from the system, which is usually done by a thermostatic element or air vent in the top of the housing.

As condensate enters the trap, the liquid level rises lifting the float and opening the valve. As this type of trap is self regulating it will operate in the modulation mode. As the condensate level rises, the outlet valve opens further handling the greater quantity of condensate. If the condensate load diminishes, the ball lowers, partially closing the valve reducing the discharge.

The valve is located below the water level, providing a water seal preventing the loss of live steam.

Float and thermostatic **(F&T)** traps are a modification of the float trap in that the thermostatic air vent is an integral part of the trap **(Figure 14.12)**.

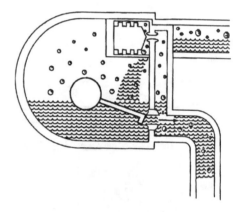

Figure 14.12, Float and Thermostatic steam trap (F&T) with thermostatic trap element in the top to bleed off air and non-condensible gases.

The thermostatic element will be open during startup, until steam enters the trap and heats the element, expanding it and closing the valve. This element will remained closed until air and gasses concentrate in the top of the trap lowering the temperature of the steam/air mixture, opening the element and discharging the mixture.

In one design of the float and thermostatic trap, there is just a "free ball" that can float up opening the orifice or sink when

Steam Traps

condensate level goes down, closing the orifice.

The failure mode for these traps is to usually have the ball float fail closed and the thermostatic element either fail open or closed.

Thermodynamic Traps

Thermodynamic traps include Disk, Piston, Lever and Orifice types.

Disc Traps

Disc traps have a single operating part, a flat disc which lifts from the valve seat to open and is forced onto the valve seat for closure.

During start up **(Figure 14.13)**, the pressure of the cold condensate and gases pushes the disc from the valve seat. This opens the trap, allowing the discharge of condensate and gasses.

chamber on the outlet side of the trap, increasing the pressure on the outlet side of the disc and decreasing the pressure below the disc, caused by high velocity flow. The combined action forces the disc onto the seat, closing the trap **(Figure 14.14)**.

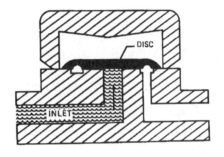

Figure 14.14, Disc trap forced closed by steam in the upper chamber.

The disc is designed to have the whole of the area of the disc exposed to pressure on the top. On the bottom, incoming side, the landing surfaces blocks pressure from acting on the whole disc area, thus unequal forces are developed. The flash steam on the top is sufficient to keep the disc seated until it cools relieving the pressure above the disc, allowing cooler condensate to flow once again.

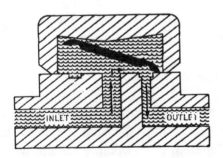

Figure 14.13, Disc trap in open position.

Discharge continues until hot condensate near steam saturation temperature enters the trap. This condensate will flash in the

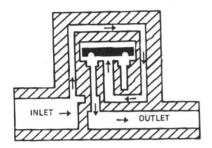

Figure 14.15, A double walled disc trap.

Orifice Traps

Orifice traps consist of one or more orifices in series and have no moving parts.

Single orifice traps typically have very small orifices in the range of one tenth the diameter of thermodynamic or thermostatic trap orifices.

The operating principle is that water (condensate) flowing through the orifice is restricted. This backs up some of the condensate preventing the flow of steam through the orifice. Steam tries to flow through the orifice at a very high velocity whereas the condensate is very slow by comparison, this phenomena serves to choke the flow of steam with condensate.

The orifice is somewhat self regulating. That is, if the condensate load is less than the orifice was designed for, the condensate will be hotter, causing more flashing, which will further restrict flow. If the condensate load is greater than the orifice was designed for, the condensate will back up and cool. This cooling will reduce flashing, thus allowing a greater condensate flow.

Orifice traps have a constant discharge, not the cyclic discharge of other thermodynamic traps.

Each orifice trap must be very closely engineered to its specific load because they have the potential to either blow steam (claimed to be an insignificant amount) or back up condensate during start up or during high loads when their capacity may be exceeded.

Thermostatic Steam Traps

Thermostatic steam traps are operated by changes in temperature and include Bellows, Diaphragm, Bimetallic and Expansive Element types. Thermostatic traps respond more slowly to changing operating conditions than some other types of traps because of the heat stored within the trap materials and the condensate which collects in the trap.

Thermostatic steam trap operating principles are simple and thermostatic steam traps usually have only one moving part.

Bellows Steam Traps

Bellows traps are often called balanced pressure or thermostatic traps because the bellows contains a volatile fluid which closely parallels the temperature-pressure relationship of the steam saturation curve. The pressure created inside the bellows by this fluid as it vaporizes from the heat of the condensate closes the trap against the pressure of the steam system **(Figure 14.16 -17)**.

Steam Traps

During start up the bellows is contracted away from the seat, allowing the discharge of condensate and gasses. Discharge continues until hot condensate enters the trap, heating the bellows and the volatile fluid. The fluid vaporizes and expands, causing the bellows to expand and close the valve.

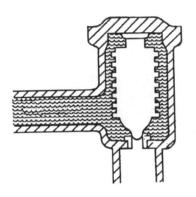

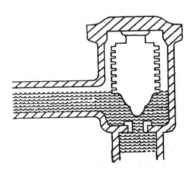

Figure 14.16, Thermostatic trap in the open position.

The trap will remain closed until the condensate, the bellows and the volatile fluid cool. This cooling condenses some of the fluid, reducing the pressure inside the bellows. This reduced internal pressure allows the external pressure, from the steam system, to contract the bellows away from the seat. This opens the trap and it begins the cycle again.

Figure 14.17, Thermostatic trap in the closed position.

Reduced temperature from a steam and non-condensible gas mixture will also cool the bellows and fluid, allowing the discharge of these gases.

Diaphragm Traps

Diaphragm traps are a modification of the bellows type trap. However, instead of a large bellows with many convolutions or welded elements, a single element is used.

Operation of this type steam trap closely parallels the operation of bellows traps.

Bimetallic Traps

Bimetallic traps **(Figure 14.18)** use the temperature of the condensate in the trap to bend bimetallic elements against the force exerted by the steam pressure on the valve. There are many different configurations of bimetallic traps which cause significant differences in operation.

Steam Traps

Various elements are used and they may be liquid or solid. In all cases, the response to changing conditions is relatively slow and dependent upon the thermal mass of the element.

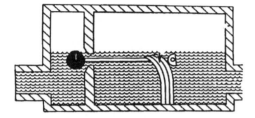

Figure 14.18, A bimetallic steam trap.

Since a bimetallic element is a nearly linear device, most manufacturers use several types of bimetal or special shapes to cause the trap to approximate the steam saturation curve over the operating range.

Because of the relatively large thermal mass of the bimetallic element, response to system changes can be slow. Additionally, these traps are sometimes affected by the back pressure which works against the opening force on the valve possibly increasing the amount of sub-cooling.

During start up the bimetallic element is relaxed, allowing the steam system pressure to open the valve discharging the cold condensate and non-condensible gases. As warm condensate reaches the trap, the element warms and changes shape, exerting a closing force on the valve.

Expansive Element Steam Traps

Expansive element traps (**Figure 14.19**) are characterized by a constant discharge temperature for a given condensate load, regardless of the steam system pressure.

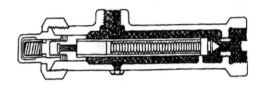

Figure 14.19, An expansive element steam trap.

During start up the element is contracted, allowing the discharge of cold condensate, air and non condensable gases. Discharge continues until warmer condensate heats and expands the element, moving the valve towards the seat. The trap will reach equilibrium condition, discharging condensate continuously at a nearly constant temperature. Only rarely will these traps cycle and then only under very light load conditions, before returning to the modulating mode.

The operation of these traps is regulated by the condensate temperature and are suitable for applications where condensate backup and a slow response to load changes is acceptable.

Steam Trap Losses

It is clear, that steam leaks are quite expensive. The high and low estimates in **Table 14.1** take into account the throttling effect of condensate choking the full flow of escaping steam with variations in condensate formation load.

Trap Orifice Diameter	High Estimate ($)	Low Estimate ($)
1/8"	3,175	2,300
1/4"	16,600	9,250
1/2"	51,000	37,000

Table 14.1 Annual cost of steam loss from traps for 100 psi steam at $5.00 per 1000 pounds.

We can assume that the cost of steam leaks through failed traps will range from $2,000 to $50,000 per trap over the course of a year.

The economic incentive for eliminating failed steam traps clearly exists. A trap anywhere in a facility may fail open at any time and the losses, if it is not found and repaired quickly, can be very large. Failure could occur at any time, even the day after the last inspection. **Figure 14.20** shows how flash steam and steam from failed steam traps vents to the atmosphere.

To determine how often traps will fail has so many variables, that any conclusion should be based on actual conditions at a specific facility. Some generalizations have emerged over the years. There is very little information on this topic.

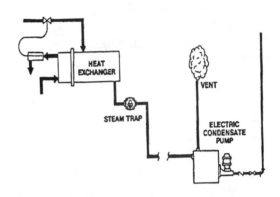

Figure 14.20, Showing heat exchanger and steam venting to atmosphere from condensate tank.

Sampling of Published Comments on Steam Trap Performance

A review of various articles published by leading authorities reveals that trap failures range from 5% to 50% of the trap population at any given facility.

"The failed trap percentage at any given time should be in the 5% to 10% range. Performance better than that is difficult in the average plant. Failed-trap percentages as high as 50% in some plants showed up during the early days of the energy crisis, and occasionally we see such performance still."

Power Magazine
April, 1980

"Experience indicates, that in plants without planned steam trap maintenance, between 10% and 50% of the traps are malfunctioning at any given time - as a result of errors in sizing, misapplication or inadequate maintenance."

Plant Engineering
March 5, 1981

"Most plants can save 10-20% of fuel cost simply by having a formal, active steam trap program...For the first year, a return of $1 million in energy savings for each $300,000 spent upgrading the system is the rule rather than the exception."

Chemical Engineering
February 9, 1981

"Trap life should be 4-5 years average. For a plant with 2000 traps this means 400-500 replacements every year."

Power Magazine
April 1980

"The average disc trap should last six months to a year...disc traps sometimes failed 'within days of installation'..."

Energy User News
June 6, 1983

"Of 260,000 installed traps studied (in 40 industrial plants) the average performance level was found to be only 58%. In a typical plant with 2,000 traps, 840 were failed; 42% needed corrective maintenance that plant personnel were not providing... inefficiencies in the energy/management area were costing the average industrial plant over $2000 in steam each day - even though $500 or more is spent daily on steam trap maintenance."

Industry Week
April 16, 1984

This review of some of the most prestigious and informed publications on efficient plant management are all of one voice. There is tremendous potential to waste energy in steam systems, also there is the same level of opportunity to correct the situation.

Steam Trap Selection Sizing and Installation

Steam Trap Piping

In most plants, the removal of condensate and air from steam systems is accomplished through steam trap stations. **Figure 14.21** shows the piping configuration for a steam trap station.

Steam Traps

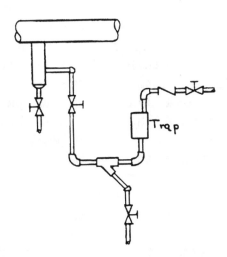

Figure 14.21, A typical steam trap station.

The first part of the steam trap station is a reservoir that connects to the lowest point of the steam line or process equipment. The reservoir is sized to collect the large quantity of condensate formed during the warm up phase when the system is brought up from the cold state. At the bottom of the reservoir is a dirt pocket and valve to collect larger particles of scale and dirt in the system. It has a blowoff valve for clean-out.

The condensate flows from the collection point through a strainer and trap to the condensate system. The station is configured with inlet and outlet valves to isolate the trap if repairs are needed.

The test connection can be opened, coordinated with the closing of the trap outlet valve to see what is coming through the trap, steam or condensate. This is one of the best ways to test trap performance.

The "Y" strainer can be cleaned out by opening the blowoff valve.

The bypass line (not shown) enables plant personnel to bypass the entire steam trap station if necessary. Bypass valves are usually opened when there is a suspected problem with a trap or to assist with drainage during cold startup. Bypassing a steam trap may also be necessary to deal with a waterlogging problem caused by a trap malfunction.

With the bypass open, steam will be wasted and the efficiency of the plant reduced. In some cases a trap is installed in the bypass line. Then, if the main trap fails, the bypass trap can be put into service and the problem of blowing steam through the bypass is eliminated.

Selecting the Type of Trap

Properly selected, sized and installed traps are the best guarantee for efficient operation, long service life and minimum downtime.

The major considerations in selecting the type of steam trap are:

A. Type of service

- Continuous or intermittent removal of condensate

- Temperature of the condensate and system pressure

- Range of load on the trap

Steam Traps

- Rate of change of the load

B. Operational

- Normal steam loss during operation

- Reliability

- Failure mode most likely to occur

The above considerations were used in developing **Table 14.2**, the Steam Trap Selection Guide. This guide was developed by the U.S. Navy for their many and varied shore bases including hospitals, industrial facilities, air bases and training centers.

Operating characteristics of various types of traps are summarized in **Table 14.3**, which gives a comparison of various common trap configurations for specified service.

When installing and evaluating traps, an understanding of their limitations is essential to long, reliable and trouble free service. The limitations listed in **Table 14.4** should be given careful consideration in the steam trap selection process.

Sizing Steam Traps

Factors that affect the accuracy of trap sizing are:

A. Estimating the maximum condensate load.

B. Range of operating pressure and differential pressure.

C. Selection of a safety factor.

Condensate Load

The amount of condensate generated by equipment can generally be obtained from equipment manufacturer's literature or specification sheets. Steam trap manufacturers, through years of experience, have also developed formulas, tables and graphs for estimating condensate load for most applications. **Table 14.5** gives samples of simplified estimating aids. **Table 14.6** gives estimates of condensate loads for various sizes of steam mains and different pressures.

Pressure differential

Trap capacity is affected by the differential pressure across a trap. If a trap exhausts to atmosphere, the differential pressure will be the supply pressure. In some plants, traps are installed with the outlet connected to a pressurized return system. The trap must operate against this head plus any requirement to lift the condensate to the return system. **Table 14.7** gives examples of the reduction in trap capacity caused by this back pressure which must be taken into account when sizing traps.

Safety Factor

The safety factor is a multiplier applied to the estimated condensate load since trap ratings are based on maximum discharge capacities or continuous flow ratings (**Figure 14.22**). Safety factors vary from 2:1 to 10:1 and are influenced by the operational characteristics of the trap, accuracy of estimated condensate load, differential pressure and expected changes in pressure and the configuration of the installation design.

278

Steam Trap Selection Guide

Application	Special Considerations	Primary Choice	Alternate Choice
Steam Mains and Branch Lines	• Energy conservation • Response to slugs of condensate • Ability to handle dirt • Variable load response • Ability to vent gases • Failure mode (open)	Inverted Bucket Thermostatic in locations where freezing may occur	Float and Thermostatic
Steam Separators	• Energy conservation • Variable load response • Response to slugs of condensate • Ability to vent gas • Ability to handle dirt • Failure mode (open)	Inverted bucket (large vent)	Float and Thermostatic Thermostatic (above 125 psig)
Unit Heaters and Air Handling Units	• Energy conservation • Resistance to wear • Resistance to hydraulic shock • Ability to purge system • Ability to handle dirt	Inverted Bucket (constant pressure) Float and Thermostatic (variable pressure)	Float and Thermostatic (constant pressure) Thermodynamic (variable pressure)
Finned Radiation and Pipe Coils	• Energy conservation • Resistance to wear • Resistance to hydraulic shock • Ability to purge system • Ability to handle dirt	Thermostatic (constant pressure) Float and Thermostatic (variable pressure)	Thermostatic

Table 14.2 Steam Trap Selection Guide

279

Steam Trap Selection Guide (continued)

Application	Special Considerations	Primary Choice	Alternate Choice
Tracer Lines	• Method of operation • Energy conservation • Resistance to wear • Variable load performance • Resistance to freezing • Ability to handle dirt • Back pressure performance	Thermostatic	Thermostatic
Shell and Tube Heat Exchangers	• Back pressure performance • Gas venting • Failure mode (open) • Resistance to wear • Resistance to hydraulic shock • Ability to purge system • Ability to handle dirt • Ability to vent gasses at low pressure • Energy conservation	Inverted bucket large vent (constant pressure) Float and thermostatic (variable pressure)	Thermostatic
Process Air Heaters	• Energy conservation • Ability to vent gases • Ability to purge system • Operation against back pressure • Response to slugs of condensate • Method of operation	Inverted bucket	Float and Thermostatic Thermodynamic
Steam Kettles: Gravity drain	• Energy conservation • Resistance to wear • Resistance to hydraulic shock • Ability to purge system • Ability to handle dirt	Inverted bucket	Thermostatic
Siphon drain	• Energy conservation • Resistance to hydraulic shock • Ability to vent air at low pressure • Ability to handle air start up loads • Ability to handle dirt • Ability to purge system • Ability to handle flash steam	Thermostatic	Thermostatic

Table 14.2 Steam Trap Selection Guide (continued)

Steam Trap Operating Characteristics

Characteristics	Bellows Thermostatic	Bimetallic Thermostatic	Disk	F & T	Inverted Bucket
● Method of operation (discharge)	Continuous (1)	Self-modulating	Intermittent	Continuous	Intermittent
● Operates against back pressure	Excellent	Poor	Poor	Excellent	Excellent
● Venting capability	Excellent	Excellent	Good (3)	Excellent	Fair
● Load change response	Good	Fair	Poor to Good	Excellent	Good
● Freeze resistance	Excellent	Excellent	Good	Poor	Poor (5)
● Waterhammer resistance	Poor	Excellent	Excellent	Poor	Excellent
● Handles start up loads	Excellent	Fair	Poor	Excellent	Fair
● Suitable for superheat	Yes	Yes	Yes	No	No
● Condensate subcooling	5 - 30°F	50 - 100°F	Steam Temperature	Steam Temperature	Steam Temperature
● Usual failure mode	Closed (2)	Open	Open (4)	Closed	Open

1. Can be intermittent on low loads.
2. Can fail open due to wear.
3. Not recommended for very low pressure.

4. Can fail closed due to dirt.
5. May be insulated for excellent resistance.

Table 14.3, Steam Trap Operating Characteristics

Steam Trap Limitation Guide

A. Bucket trap

- Trap will not operate where a continuous water seal cannot be maintained.

- Must be protected from freezing.

- Air handling capacity not as great as other type traps.

B. Ball float trap

- Must be protected from freezing.

- Operation of some models may be affected by waterhammer.

C. Disk trap

- Not suitable for pressures below 10 psi.

- Not recommended for back pressures greater than 50 percent of inlet pressure.

- Freeze proof when installed as recommended by the manufacturer.

D. Impulse orifice trap

- Not recommended for back pressures greater than 50 percent of inlet pressure.

- Not recommended where subcooling condensate 30°F below the saturated steam pressure is not permitted.

E. Thermostatic trap

- Limited to applications in which the condensate can be held back and subcooled before being discharged.

- Operation of some models may be affected by waterhammer.

F. Combination float and thermostatic trap.

- Cannot be used on superheated steam systems.

- Must be protected from freezing.

- Operation of some models may be affected by waterhammer.

Table 14.4, Trap Limitations

General Formulas for Estimating Condensate Loads

Application	Lb/Hr of Condensate
Heating Water	$= \dfrac{\text{GPM}}{2} \times \text{Temperature Rise } °F$
Heating Fuel Oil	$= \dfrac{\text{GPM}}{4} \times \text{Temperature Rise } °F$
Heating air with Steam coils	$= \dfrac{\text{CFM}}{900} \times \text{Temperature Rise } °F$
Heating: pipe coils and radiation	$= \dfrac{A \times U \times (\text{Steam } °F - \text{Air } °F)}{L}$

A = Area of heating surface
U = Heat transfer coefficient
    (2 for free convection)
Delta T = Steam temperature - Air temperature, °&F
L = Latent heat of steam BTU/Lb

Table 14.5, General formulas for estimating condensate loads.

283

Estimating Condensate Load for Insulated Steam Mains

Insulated Steam Main

Lb/Hr of condensate per 100 lineal foot
at 70°F (at 0°F, multiply by 1.5)

Steam Pressure (psig)	Size of Main Inches					
	2	4	6	8	10	12
10	6	12	16	20	24	30
30	10	18	25	32	40	46
60	13	22	32	41	51	58
125	17	30	44	55	68	80
300	25	46	64	83	103	122
600	37	68	95	124	154	182

Table 14.6, Estimating condensate load for insulated steam mains.

Inlet Pressure (psig)	Back Pressure (% of Inlet Pressure)		
	25%	50%	75%
10	5%	18%	36%
30	3%	12%	30%
100	0	10%	28%
200	0	5%	23%

Table 14.7, Percentage reduction in steam trap capacity

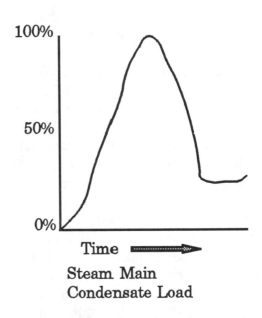

Steam Main
Condensate Load

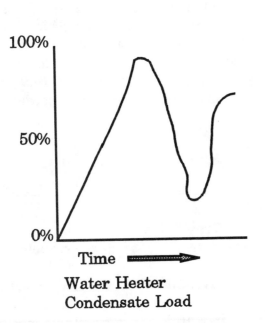

Water Heater
Condensate Load

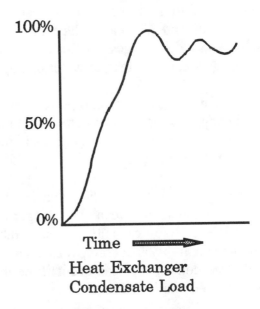

Heat Exchanger
Condensate Load

Figure 14.22, Condensate load change over time on a cold start up for a water heater, steam main and heat exchanger.

Steam Traps

Safety factors are needed to cope with the varying condensate loads experienced in the "real world " of actual use.

During start up conditions the cold piping will generate a large amount of condensate until it is up to operating temperature. Whether this occurs over several hours or a few minutes the amount of condensate will be the same, but the trap will have to handle it more quickly requiring added capacity.

To add to the challenge of condensate removal during warm up, the system pressure is lower than normal so the trap output will be lower than rated capacity. Waterhammer must be avoided so the trap must be designed for this unfavorable condition.

Other conditions affect condensate load also, cold winds and rain can reduce insulation values. Sudden surges in load on heat exchangers can also lead to waterlogging.

A boiler which has carry-over or is producing low quality steam containing unevaporated water places an unusually high load on the whole system.

Waterlogging cuts production and comfort and waterhammer can be dangerous. Insufficient trap capacity causes these problems. On the other hand, wasted energy and large dollar losses can occur if the trap is oversized and is leaking.

The answer to this situation may be an effective trap maintenance program.

Steam Trap Maintenance

The cost of steam lost through a leaking trap in one day can exceed the cost of a new trap. If a trap is backing up condensate and has become waterlogged, the reduction in capacity or efficiency of steam-using equipment can also waste more money in one day than the cost of the trap.

Good inspections that pinpoint problem traps for repair or replacement are important. While the trap is the focus of tests and inspections, the steam trap station and its piping and fittings should be checked also as these things may be contributing to trap problems. Missing strainers, open bypass lines and condensate piping pressure are a few problems that could be found.

Inspection schedules

The ideal is all traps functioning properly. This must be balanced against cost effective use of time and materials. Each location offers different challenges. Dirty systems, above normal acid accumulations and problems with the trapping stations all point to judgment playing a major role in trap maintenance. **Table 14.8 and 14.9** are commonly used trap inspection schedules.

Trap Failure Rate	Inspection Frequency
Over 10%	Two months
5 - 10%	Three months
Less than 5%	Six months

Table 14.8, Trap inspection schedule.

Steam Traps

Pressure (psi)	Inspection Frequency
0 - 30	Annual
30 - 100	Semiannual
100 - 250	Quarterly or monthly
Over 250	Monthly or weekly

Table 14.9, One manufacturer recommends inspection frequencies based on system pressure.

Traps serving critical process equipment may need special inspections during very cold weather to guard against freezing damage.

The basic methods of inspecting traps are:

    A. Visual observation
    B. Sound detection
    C. Temperature measurements.

Visual observation is the best and least costly method of checking trap operating conditions, but none of the methods provide a cure-all for trap troubleshooting.

Table 14.10 is a steam trap troubleshooting guide, useful for trap inspections.

Visual Observation

Observing the discharge from a trap is the only positive way of checking its operation.

No special equipment is required but training and experience are necessary, particularly for recognizing the difference between flash steam and live steam.

- Flash steam is the lazy vapor formed when the hot condensate gives up energy when reducing pressure through the trap.

- Live steam is a higher temperature, higher velocity discharge, and usually leaves the discharge pipe with a short clear, almost invisible, section of flow before condensation begins and a visible steam cloud develops.

Sound Detection

Some traps can be heard cycling on and off as they discharge condensate. The disk, inverted bucket and piston type of traps have this characteristic.

Some traps continually modulate like the float traps or expansive element and some thermostatic traps; they only give off flow sounds. However if they aren't working, they don't give off any sound.

An automotive stethoscope or other simple sound transmission device may be adequate. If there is a lot of noise around or a lot of traps giving off noise, then an ultrasonic listening device may be warranted. They can sense high frequency (an indication of flow) and low frequency mechanical noises (indicates operation).

Temperature Measurements

Diagnosing trap condition from temperature differences between upstream and downstream pipes is the least reliable

Steam Trap Troubleshooting Guide

Steam Traps

TRAP TYPE	NORMAL OPERATION	PROBLEM INDICATION	POSSIBLE CAUSE
		VISUAL INSPECTION	
ALL TYPES	• Trap hot under operating conditions • Discharge a mixture of condensate and Flash steam. • Cycling open/closed depending on type of trap. • Relatively high inlet temperature	• Live steam discharge, little entrained liquid • Condensate cool, little flash steam • No discharge • Leaking steam at trap	Failed open Holding back condensate. Failed closed; clogged strainer: line obstruction Faulty gasket
		TEMPERATURE MEASUREMENT	
		• High temperature down stream • Low temperature upstream	Failed open Failed closed; clogged strainer; line obstruction
		SOUND INSPECTION	
FLOAT AND F & T	• Continuous discharge on normal loads may be intermittent on light load • Constant low pitched sound of continuous flow	• Noisy high pitched sound • No sound	Steam flowing through; Failed open. Failed closed
THERMOSTATIC	• Discharge continuous or intermittent depending on load, pressure, type. • Constant low pitched sound of continuous or modulating flow.	• Same as for Float, F & T above.	
INVERTED BUCKET	• Cycling sound of bucket opening and closing • Quiet steady bubbling on light load	• Steam blowing through • No sound • Discharging steadily; no bucket sound. • Discharging steadily;bucket dancing • Discharging steadily; bucket dancing after priming • Discharging steadily; no bucket sound.	Failed open Failed closed Handling air, check later Lost prime Failure of internal parts. Trap undersized
DISK	• Intermittent discharge • Opening and snap closing of disk about every 10 seconds	• Cycles faster than every 5 seconds. • Disk chattering over 60 times/min or no sound	Trap undersized or faulty. Failed open

Table 14.10, Troubleshooting Steam Traps

288

inspection method. It can be useful in combination with visual or sound inspection as long as the potential ambiguities are recognized.

There is a wide range of temperature measurement equipment. It ranges from infrared devices, which are handy for reading temperatures from a distance and at inaccessible locations. Standard pyrometers and surface thermocouples are also suitable. Heat sensitive color markers are also used at some locations.

Take temperature measurements immediately adjacent to, not more than two feet away, either side of the trap. Temperature readings should be in the ranges shown in **Table 14.11** for the pressures in the in and out lines. For example, for a steam system with a pressure of 150 psi, if the trap inlet temperature is 340°F and the outlet side, with a pressure of 15 psi, has a temperature of 230°F this indicates expected temperature conditions.

Table 14.12 is a steam trap inspection check list to aid personnel in their trap checking routines. They should also investigate the possibility of trap misapplication. **Table 14.13** will be helpful for this purpose.

One manufacturer has developed an automatic system for detecting if steam traps are blowing-off an excessive amount of steam. The sensor chamber **(Figure 14.23)** has a small orifice in a division plate which is designed to pass "normal" steam flow. If this flow increases substantially, the level goes down on the upstream chamber exposing a sensor which signals failure of the trap **(Figure 14.24)**.

Steam Pressure (psig)	Pipe Surface Temperature Range (°F)
0 (Atmosphere)	212
15	225 - 238
30	245 - 260
100	305 - 320
150	330 - 350
200	350 - 370
450	415 - 435
600	435 - 465

Table 14.11, Normal pipe temperatures at various operating pressures.

Steam Trap Inspection
Check off List

FOR ALL TRAPS

- Is steam on?
- Is trap hot - at operating temperature?
- Wet test for signs of a hot trap; squirt a few drops of water on trap. Water should start to vaporize or sizzle immediately. If it does not this indicates a cold trap.
- Tag traps for maintenance check to see if this is a trap or system problem.
- Blowdown strainer.

SOUND CHECK HOT TRAP

- Listen to trap operate.
- Check for continuous flow:
    - Low pitch condensate flow
    - High pitch steam flow
- Check for intermittent flow
- Is trap cycling?
- Note mechanical sounds.

VISUAL CHECK TRAPS THAT SOUND BAD

- Close valve return line.
- Open discharge valve.
- Observe discharge for:
    - Normal condensate and flash steam
    - Live steam
    - Continuous or intermittent operation

TEMPERATURE CHECK IF NECESSARY

- Clean spots upstream and downstream of trap for measuring temperature.
- Record supply line pressure.
- Measure supply line temperature.
- Record trap discharge line pressure.
- Measure trap discharge line temperature.
- Tag failed traps for replacement or shop repair.

CHECK EXTERNAL CONDITIONS

- Supports and braces
- Insulation
- Corrosion
- leaks

Table 14.12, Steam trap inspection check off list.

Inspection List
for
Trap Misapplication

1. Is trap installed backwards or upside down?

2. Trap located too far away from the equipment being serviced.
   Piping runs too long.

3. Traps not installed at low points or sufficiently below steam-using equipment to insure proper drainage.

4. Traps oversized for conditions. Oversized traps may allow steam to blow through.

5. More than one item of equipment served by one trap. "Group trapping" is likely to short circuit one line due to differences in pressure and other lines will not be properly drained.

6. The absence of check valves, strainers and blowdown cocks where required for efficient operation.

7. Trap vibration due to insecure mounting.

8. Bypass line with open valves. Bypass should be fitted with a standby trap.

9. Condensate line elevation higher than steam pressure through trap can lift. Inlet steam pressure must be high enough to lift condensate to drain system.

10. Inverted bucket and thermostatic traps that may be exposed to freezing. Freezing may affect operation of other traps too.

11. Thermostatic and disk traps must give off heat to operate, check to see that these types are not insulated.

12. Disk trap with excessive back pressure may not have enough differential pressure to operate properly.

Table 14.13, Inspection check list for trap misapplication.

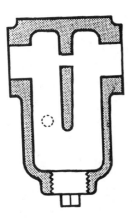

Figure 14.23, sensor chamber for trap monitoring installation. Note the orifice and weir (partition plate).

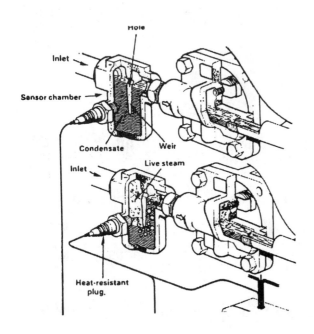

Figure 14.24, trap monitoring installation.

Chapter 15

Boiler Water Treatment

Impurities in Boiler Feed Water Concentrate in the Boiler

All boiler water contains dissolved solids. When the feed water is heated, it evaporates and goes off as distilled steam leaving impurities behind. As more and more water is distilled in the boiler, more feed water is added to replace it. As a result the amount of these solids dissolved in the boiler water gradually increases. After a while there is so much of these highly soluble solids in solution in the boiler water that it does not boil like ordinary water, it boils more like a syrup.

The bubbles of steam which rise to the surface of the water do not readily break free from the surface. Instead big bubbles form. When they break, they carry with them into the steam space some of the film that formed the bubble material. This condition is commonly referred to as carry-over.

In addition some highly soluble materials are changed by the high boiler temperature to materials of low solubility such as calcium carbonate which are then precipitated. Much of this precipitation takes place where the boiler water is the hottest, where water is in contact with high heat transfer zones. The precipitate is deposited on the heating surface and forms a scale build up.

This scale is a good insulator, reducing heat transfer. As scale builds up, the steam and water is unable to keep the tube metal surfaces cool and it begins to overheat. This overheating destroys the strength of the tubing causing tube failure. This can occur as blistering which ruptures or general melting, depending on the circumstances.

One of the purpose of a water treatment program is to keep certain scale forming solids in solution. Other scale forming solids are turned into a soft fluffy precipitate and carried down to the low points of the boiler. **Table 15.1** shows the effect of improper or inadequate water conditioning.

Cycles of Concentration

All of the impurities dissolved in water are usually termed Total Dissolved Solids, referred to as TDS. Modern methods utilize electronic instruments to measure the conductance, the opposite of resistance, of boiler water. These readings are called "mhos" or "micromhos" and can be mathematically converted to parts per million with respect to sodium ions by simply using a multiplier.

One part per million (ppm) is one pound in a million pounds of water. Since water weighs approximately 8.33 pounds per gallon, one ppm is one pound in 120 thousand gallons of water.

If a given water had a total dissolved solids of 500 ppm and we concentrated this water two times or two cycles, the TDS level would be 1,000 ppm. At three cycles the TDS would be 1,500 ppm and at four cycles 2,000 ppm and so on.

In the case of a boiler as small as 100 horsepower, it can evaporate more than 10,000 gallons of water in 24 hours. If this

Boiler Water Treatment

Effects of Inadequate or Improper Water Conditioning

Effect	Problem	Remarks
Scale	Silica	Forms a hard glassy coating on internal surfaces of boiler. Vaporizes in high pressure boilers and deposits on turbine blades.
	Hardness	$CaSO_4$, $MgSO_3$, $CaCO_3$ and $MgCO_3$ form scale on boiler tubes.
Reduced heat transfer	Scale & Sludge deposits	Loss of efficiency, wasted fuel.
Corrosion	Oxygen	Causes pitting of metal in boilers and steam and condensate piping.
	Carbon Dioxide	Major cause of deterioration of condensate return lines.
	Oxygen + Carbon Dioxide	Combination is more corrosive than either by itself.
Foaming & Priming	High boiler water concentrations	Distribution system contamination, wet steam and deposits in piping, on turbine blades and valve seats.
Caustic embrittlement	High caustic concentrations	Causes intercrystalline cracking of boiler metal.
Economic losses	Repair of boilers	Repair pitted boilers and clean heavily scaled boilers.
	Outages	Reduce efficiency and capacity of plant.

Table 15.1. Effect of inadequate or improper water conditioning.

Boiler Water Treatment

water had a hardness of 340 ppm, 28 pounds of residue would be left behind in the boiler every day.

Energy loss through blowdown is minimized by maintaining the boiler water cycles of concentration as close to the recommended limit as possible. This can be best accomplished by automating the continuous blowdown. **Figure 15.1** shows how blowdown losses can be reduced by increasing cycles of concentration of boiler water.

Automatic control of the continuous blowdown involves the use of a continuous conductivity monitor to activate the blowdown control valve. These units can maintain boiler water conductivity within close limits and thereby minimize excessive blowdown loss, which occurs with manual control.

Water Hardness

Hardness in the boiler water indicates the presence of relatively insoluble impurities. Impurities may be classed as (a) dissolved solids, (b) dissolved gasses and (c) suspended matter. Actually, in the heating and concentration of boiler water, these impurities will precipitate even more rapidly since they are even less soluble at high temperatures. Waters containing large amounts of calcium and magnesium minerals are "hard to wash with." The amount of hardness in normal water may vary from several ppm to more than 500 ppm. Because calcium and magnesium compounds are relatively insoluble in water, they tend to precipitate out, causing scale and deposit problems. The hardness of the water source is an important consideration in determining the suitability of water for steam generation.

As mentioned earlier, this process of precipitation will take place on the heat exchange surfaces and is known as scale.

Condensate System Corrosion

The most prevalent type of condensate system corrosion is caused by carbon dioxide.

Carbon dioxide enters the system with the boiler feed water in the form of bicarbonate and carbonate alkalinity. When exposed to boiler temperatures, bicarbonates and carbonates break down to form carbon dioxide. The carbon dioxide is carried away in the steam and is condensed to form carbonic acid.

The Priming Boiler

Priming occurs when slugs of water enter the steam distribution system. There are a number of causes such as: high steam demand, large and sudden drops in system pressure, blowdown, quick opening valves in the distribution system and unsuitable steam nozzle and steam header size.

One way priming could happen is after an operator has brought a boiler up to operating pressure he opens the distribution header quickly. Then, within a few minutes, he notices the water bouncing up and down in the sight glass. This bouncing can become quite violent and the boiler can shut down on low water cut off.

What has happened? The piping and heat exchange equipment serviced by the distribution header is cold and the steam is quickly condensed possibly forming a vacuum. This causes a violent boiling action in the boiler, it could even start a series of

295

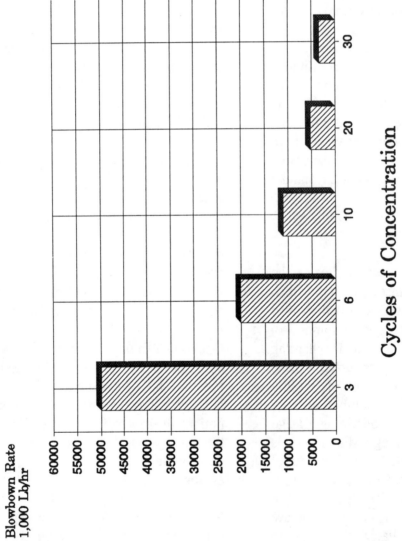

Figure 15.1. Blowdown is reduced dramatically when Cycles of Concentration is increased.

Boiler Water Treatment

oscillations where the water surface rises or "mounds" towards the steam nozzle. The lift of the water can not be sustained and it drops starting a sloshing effect. A wave or bouncing effect can start which will allow water to enter the steam system in slugs; a dangerous condition. This condition can also upset boiler water circulation patterns.

One rule of thumb used in raising boiler water temperature is to go at a rate of 100 F per hour. Open valves slowly, allowing the system to warm up and balance out before introducing sudden surges on the system.

Deareators

Oxygen and carbon dioxide are very harmful to boiler systems. Deareators are designed to remove dissolved gases from the boiler feed water. They are effective and oxygen can be reduced to trace levels; about .005 ppm. While deareators are efficient, traces of oxygen can cause a significant amount of corrosion, so chemical treatment is also used.

While the deareator removes carbon dioxide from the feed water, bicarbonate and carbonate alkalinity in the boiler will produce additional carbon dioxide. This will require some water treatment in the boiler.

A deareator usually consists of a heating and deareating section. The storage section of these units are often designed to hold about 10 minutes of rated capacity of boiler feed water.

The water enters the deareator and is broken into a spray or mist and scrubbed with steam to force out the dissolved gasses.

Steam and non-condensibles flow upward into the vent condensing section where the steam is condensed. The released gases are discharged to atmosphere through the vent outlet.

Continuous Blowdown Heat Recovery

The continuous blowdown, sometimes called the surface or skimmer blowdown, is most effective in controlling the concentration in boiler water. Where continuous blowdown systems are used, the bottom blowdown is used for removal of precipitated impurities, especially those which tend to settle in the lower parts of the boiler.

Heat exchangers can be used with the continuous blowdown to recover energy from the boiler water which is being expelled from the boiler.

How pure must feedwater be?

Feedwater purity is a matter of both quantity of impurities and the nature of the impurities. Some impurities such as hardness, iron and silica for example are more concern than sodium salts. The purity requirements for any feedwater depend on how much feedwater is used as well as what the particular boiler design is. Pressure, heat transfer rate and operating equipment on the system such as turbines have a lot to do with feedwater purity.

Feedwater purity requirements can vary widely. A low pressure firetube boiler can usually tolerate higher feedwater hardness, with proper chemical treatment, while virtually all impurities must be removed from the water used with most modern high pressure watertube boilers.

Boiler Water Treatment

Water Carryover in Steam

The water evaporated to produce steam should not contain any contaminating materials, however there will be water droplets carried into steam due to several processes.

Mist Carryover

A fine mist is developed as water boils. This process is illustrated in **Figure 15.2**. A bubble of water vapor (steam) reaches the water surface and bursts, leaving a dent in the water. Liquid collapses in on the dent, with the center rising at a faster rate than the edges. This results in a small droplets being tossed free of the boiler water surface. These droplets form a fine mist. This mist is removed to a great extent in the dry portion of the boiler. However, any mist that remains entrained in the steam will have the same level of contamination as the boiler water.

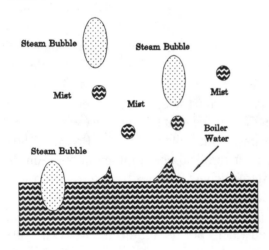

Figure 15.2. Mist formation in boiler water with high impurity levels. The volume above the water is steam, steam bubble is shown for illustration.

Foaming Carryover

The alkalinity, TDS and suspended solids can interact to create a foam in the boiler. A light foam will reduce the problem of misting to some extent. A heavy foam layer is another source of liquid carryover into the steam. The level of foaming can normally be controlled to a reasonable level by maintaining the total alkalinity at a level less than 20 percent of the TDS and maintaining the total suspended solids at a level less than 8 percent of the TDS. Antifoam agents are added to the sodium lignosulfonate sludge dispersant to help control foaming.

Priming Carryover

Priming carryover is caused by liquid surges into the steam drum that throw water into the steam space where it is carried into the steam header. Priming is always caused by a mechanical problem or mechanical properties such as oversensitive feedwater controls or incorrect blowdown procedures. There is no chemical control method available.

Silica Carryover

The silica in the boiler water can evaporate and enter the steam, independent of water carryover. The silica can form a deposit on turbine blades and other equipment when the steam condenses. This problem is controlled by maintaining a low silica level in the boiler water. The suggested limits are shown in **Table 15.2**.

Boiler Water Treatment

Boiler Pressure (psig)	Allowable Silica (as SiO_2)
0 - 15	200
16 - 149	200
150 - 299	150
300 - 449	90
450 - 599	40
600 - 749	30
750	20

Table 15.2, Silica limits in boiler water.

Determining the Amount of Carryover

The best indication of carryover is a measurement of steam conductivity. A steam conductivity of 20 to 30 micromho indicates there is a small chance that carryover is significant.

A High Conductivity Measurement Indicates Carryover

A high conductivity measurement in the steam <u>condensate</u> means there is either carryover or leakage into the steam system. The hardness must also be checked in this case. If any hardness is found, then the contamination of the condensate indicated by the high conductivity is due to leakage into the condensate system rather than carryover. This is because the carryover of boiler water should be at a very low hardness due to chemical treatment or the very low makeup water requirement.

Removal of Oxygen From Feedwater

Mechanical removal of oxygen from feedwater requires a deareating heater in which the makeup water and condensate return are scrubbed with live steam using direct contact with trays, packing, sprays or any combination thereof. This steam scrubbing boils the oxygen and other noncondensible gases out of the water. The oxygen is vented with a small amount of steam.

Two Key Operating Controls For Deareators

There are two key operating controls for deareators that must be watched. First, the deareator vent must be checked to see that a plume of steam is always flowing. Second, the pressure of the deareator and temperature of the outlet water must be controlled. At a given pressure the temperature should be within 2°F of the temperatures shown in **Table 15.3**, based on the elevation of the boiler site. If there is low or no steam flow or a low water temperature, the deareator is not operating properly.

A mixture of oxygen and water is a very corrosive combination. This corrosivity doubles with every 18°F increase in temperature.

Oxygen corrosion can be recognized by pits, typically found in the top of the steam drum or at the waterline. Oxygen can be removed from the feedwater by mechanical or chemical deareation; a combination of these methods is commonly used.

Boiler Water Treatment

Deareator Pressure (psig)	Deareator Water Outlet Temperature (°F)
0	212
1	215.3
2	218.5
3	221.5
4	224.4
5	227.1
6	229.8
7	232.2
8	234.8
9	237.1
10	239.4
11	241.6
12	244.4
13	246.4
14	248.4
15	250.3
16	252.2
17	254.1
18	255.3
19	257.0
20	258.8

Table 15.3, Deareator water outlet temperature for boiler systems at various pressures for sea level.

Chemical Removal of Oxygen From Boiler Feedwater

A mechanical deareator can reduce the oxygen content of feedwater to a fraction of a ppm. Complete removal requires additional chemical treatment. One process used is catalyzed sodium sulfite.

The chemical reaction with sodium sulfite will consume 7.88 pounds of pure sodium sulfite with one pound of oxygen. In practice, about 10 pounds per pound of oxygen are added to carry a small excess of sulfite in the boiler water. The excess that should be carried is based on the boiler pressure, according to the levels shown in **Table 15.4**.

Boiler (psig)	Sulfite Residual (as ppm SO_3)
0 - 15	30 - 60
16 - 149	30 - 60
150 - 299	30 - 60
300 - 449	20 - 40
450 - 599	20 - 40
600 - 749	15 - 30
750 -	15 - 30

Table 15.4, Boiler water sulfite levels.

Higher sulfite levels can be wasteful. In addition, sulfite can break down and cause steam and condensate to become corrosive. It should be added continuously.

Condensate Corrosion

Any free carbon dioxide in the feedwater should be removed by the deareator. However carbon dioxide in the combined form can enter the boiler as carbonates and bicarbonates in the feedwater. Under the influence of heat and pressure, this combined carbon dioxide will form free carbon dioxide which will leave the boiler with the steam.

Boiler Water Treatment

Controlling CO₂ With Neutralizing Amides

Carbon dioxide in all steam boiler systems can be controlled by neutralizing with a volatile amine. This is an amine that can be added to a boiler, where it will vaporize and pass over with the steam.

Two amides usually used are morpholine and cyclohexylamine. When steam first condenses, morpholine will be present in a larger concentration. At more distant points in the distribution system, cyclohexylamine will be available in larger concentrations and can be more effective in controlling corrosion. The two chemicals are used together for best protection.

Controlling CO₂ with Filming Amides

Another way for controlling carbon dioxide corrosion is the use of filming amides such as octadecylamine. This chemical will coat the condensate pipe and prevent the carbon dioxide in the water from coming into contact with the pipe wall. These chemicals are usually used at a level from .7 to 1.0 ppm and they are difficult to handle and mix. They must be added directly to the steam header.

What is the basis for choosing between neutralizing and filming inhibitors?

The proper choice depends on the boiler system, plant layout, operating conditions and feedwater composition. In general, volatile amines are best suited to systems with low makeup, low feedwater alkalinity and good oxygen control.

Filming inhibitors usually give more economical protection in systems with high make up, air in-leakage and high feedwater alkalinity or where the system is operated intermittently. In most cases a combination of these treatments may be best to combat condensate system corrosion.

External Water Treatment

Makeup Water

Makeup water is water added to the boiler system from an external source to replace water lost in the boiler room and in the distribution system. This includes blowdown water, steam leaks, condensate losses and steam used directly in process applications.

The usual source of makeup water is the potable water supply or what has been referred to in many cases as city water. This represents a treated water that has a predictable and uniform quality on a day-to-day basis. Other sources of makeup water include well water, surface water or holding ponds that are not treated to the extent that the potable water source is treated.

The uniformity of makeup water quality is important if the boiler water system is to be operated reliably.

Makeup water treatment varies on the needs of a particular installation. Various processes are used to improve makeup water quality including:
   a. Lime-Soda Softening
   b. Ion Exchange Process-General
   c. Sodium Ion Exchange
   d. Hydrogen Ion Exchange
   e. Deionization
   f. Dealkalization
   g. Distillation
   h. Reverse Osmosis
   i. Electrodialysis

Boiler Water Treatment

The makeup water is combined with the condensed steam returned from the distribution system (called condensate return) to become boiler feedwater. The feedwater is deareated to strip out noncondensible gases and treated with oxygen scavengers.

Internal Water Treatment

The removal of scale-forming materials from the boiler makeup by reducing the hardness to near zero is the best control method.

Internal treatment of boiler water refers to chemical additions required to prevent scale formation from materials not removed by makeup treatment and to prevent sludge deposits from forming due to the precipitation of these materials.

Deposit Formation

There are two basic causes of boiler deposit formation:

- Scale; the high temperatures found in boilers cause precipitation of compounds whose solubilities are inversely proportional to the solution temperature.

- Sludge; the concentration of boiler water causes certain compounds to exceed their maximum solubility at a given temperature, forcing precipitation in areas of highest concentration.

While these represent a somewhat simplified view of the mechanisms involved, they do summarize the factors essential to boiler deposits.

Scale

The build up of boiler scale constitutes a growth of boiler crystals on waterside heat transfer surfaces and is most severe in those areas of the boiler where maximum heat transfer occurs. **Figure 15.3** shows the relationship between heat transfer efficiency and scale deposit thickness.

Problems caused by scaling

The steam boiler normally uses an external heat source at a much higher temperature than the boiler water. The metal tubes in the boiler are kept cool by the boiler water. When scale forms, it acts like an insulation material between the water and the metal. This results in tubes operating at higher temperatures. The greater the thickness of the scale, the greater the insulating effect, and the higher the temperature of the tubes. At sufficiently high temperatures, the tube can lose tensile strength and rupture.

Sludges

Sludges are precipitated directly in the main body of the boiler water when their solubilities are exceeded. Sludge deposition usually occurs when binders are present or when water circulation is such that it allows sludge settling on hot spots allowing it to "bake" on to hot surfaces.

TDS in a Boiler

TDS in a boiler is one of the parameters used to control the water treatment program. Dissolved solids are continually added to the boiler makeup water. These solids are not evaporated with the steam; as a result the TDS becomes more concentrated as more steam is generated.

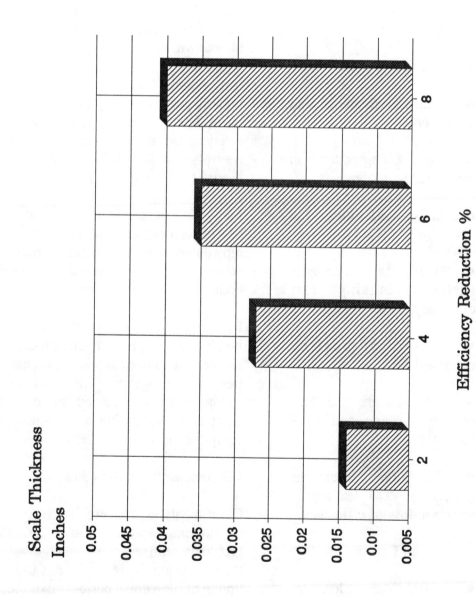

Figure 15.3. Loss of heat transfer efficiency with scale thickness.

Boiler Water Treatment

The level to which the TDS will concentrate is determined by the amount of these salts removed in the blowdown. Control of TDS level is critical in boiler operation. The higher TDS levels result in higher boiler efficiencies, but TDS levels that are too high will interfere with boiler operation.

The Consequences of Too Little Blowdown

TDS Too High

- Corrosive to boiler metal

- Causes foaming and carryover

- Alters boiling patterns in tubes leading to deposits.

Suspended Solids and Sludge too High

- Bakes onto heat transfer surfaces causing lost efficiency.

- Alters boiling characteristic

- Dirty boilers; the high cost of cleaning and extra down time.

- Absorption of chemicals (especially PO_4) leading to scale and loss of efficiency.

Hardness and Salt Levels Too High

- Scale and loss of efficiency

Alkalinity Too High

- Scale and loss of efficiency

Energy Conservation Through Optimum Blowdown Control

While blowdown is a key to safe, clean boiler operation, it must be remembered that blowdown water leaving the boiler carries a high level of Btus with it. Energy conservation requires maintaining the highest permissible cycles of concentration in the boiler water. To do this a margin of safety often must be sacrificed and good controls installed to insure that no damage occurs to the boiler as a result.

Many plants are able to increase their cycles of concentration and therefore reduce blowdown by reducing total solids concentration in the feedwater or altering the boiler water treatment program.

Reducing the solids content of the makeup involves a change in the plant makeup water source or altering the external water treatment program. The makeup water source can be changed or pretreatment equipment such as filters, deionizes or other equipment can be installed.

Efficient Bottom Blowdowns

Often plants will adhere to a strict schedule of bottom blowdown and still develop problems with excessive sludge buildup in the mud drum. This can be due to improper timing of blowdown periods. Experience has shown that frequent blowdowns of short duration (10 - 20 seconds) are more

Boiler Water Treatment

effective in removing sludge than occasional blowdowns of longer duration.

The blowdown is only effective for the first few seconds of the blowdown. Blowdowns of long duration create a great deal of turbulence in the mud drum "stirring up" the sludge level. With the sludge in suspension from this action, it can be swept up the generating tubes where it can bake onto the tube surfaces, resulting in deposits.

TDS are controlled by continuous blowdown, which is typically removed from the steam drum. Guidelines published for government plants are shown in **Table 15.5**.

Boiler Pressure (psig)	Maximum TDS (ppm)	Maximum Conductivity (μmho)
0 - 15	6,000	9,000
16 - 149	4,000	6,000
150 - 299	4,000	6,000
300 - 449	3,500	5,250
450 - 599	3,000	4,500
600 - 749	2,500	3,750
750 -	2,000	3,000

Table 15.5, Total dissolved solids (TDS) and conductivity limits for steam boilers.

Boiler Blowdown Calculations

The rate of blowdown from a boiler is a critical operating control for TDS.

a. The water added to a boiler must equal water loss from the boiler.

$$F = E + B$$

F = feedwater lb/hr
E = Steam generated, lb/hr
B = Blowdown, lb/hr

b. The blowdown can be related to the feedwater using cycles of concentration (COC).

$$C = F/B$$

C = Cycles of Concentration
F = Feedwater, lb/hr
B = Blowdown, lb/hr

Table 15.6 is the American Boiler Manufacturers Association (AMBA) specified limits for boiler water composition with respect to operating pressure to assure good quality steam.

Table 15.7 was developed by the American Society of Mechanical Engineers (ASME), Research Committee on Water in Thermal Power. It shows the need for feedwater to be extremely pure. With today's designs, heat-flux rates of 250,000 Btu/hr/sq-ft are anticipated. Combined with dimensional restrictions of modern design, this has raised the need for the limits in **Table 15.7**. These new guidelines are only suggested limits that will continue to be refined.

Maximum Limits for Boiler Water

Boiler Pressure (psig)	Total Solids (ppm)	Alkalinity (ppm)	Suspended Solids (ppm)	Silica (ppm)
0 - 300	3,500	700	300	125
301 - 450	3,000	600	250	90
451 - 600	2,500	500	150	50
601 - 750	2,000	400	100	35
751 - 900	1,500	300	60	20
901 - 1,000	1,250	250	40	8
101 - 1,500	1,000	200	20	2.5
1501 - 2,000	750	150	10	1.0
Over 2,000	500	100	5	0.5

Table 15.6. American Boiler Manufacturers Association (ABMA) limits for boiler water composition.

Boiler Feedwater

Drum Pressure (psig)	Iron ppm Fe	Copper ppm CU	Total Hardness ppm CaCO$_3$	Silica ppm SiO$_2$	Total Alkalinity[1] ppm CaCO$_3$	Specific Conductance µmho/cm
0 - 300	0.100	0.050	0.300	150	350[2]	3500
301 - 450	0.050	0.025	0.300	90	300[2]	3000
451 - 600	0.030	0.020	0.200	40	250[2]	2500
601 - 750	0.025	0.020	0.200	30	200[2]	2000
751 - 900	0.020	0.015	0.100	20	150[2]	1500
901 - 1000	0.020	0.015	0.050	8	100[2]	1000
1001 - 1500	0.010	0.010	ND[4]	2	NS[3]	150
1501 - 2000	0.010	0.010	ND[4]	1	NS[3]	100

[1]Minimum level of hydroxide alkalinity in boilers below 1000 psig must be individually specified with regard to silica solubility and other components of internal treatment.

[2]Maximum total alkalinity consistent with acceptable steam purity. If necessary, the limitation on total alkalinity should override conductance as the control parameter. If makeup is demineralized water at 600 - 1000 psig, boiler water and conductance should be as shown in the table for the 1001 - 1500 psig range.

[3]NS (Not specified) in these cases refers to free sodium - or potassium-hydroxide alkalinity. Some small variable amount of total alkalinity will be present and measurable with the assumed congruent control or volatile treatment employed at these high pressure ranges.

[4]ND is None Detectable

Table 15.7. Boiler water limits developed by the ASME Research Committee on Water in Thermal Power Systems.

Appendix A

Figures

Chapter 1 Optimizing Boiler Plants, Establishing the Ideal Scene.

Chapter 2 Boiler Plant Efficiency

Chapter 3 Distribution System Efficiency

Chapter 4 Energy Management Basics for Boilers and Systems.

Appendix A

Figures

Figure 10.8. Smoke limit vs. oxygen relationship for adjusting controls for optimum efficiency. Point (A) is minimum oxygen control point, (B) is set point based on buffer zone for various system errors (dashed line).

Figure 10.9. An oxygen trim system compensates for many control system errors.

Figure 10.10. A carbon monoxide trim system continuously seeks the optimum air/fuel ratio. In contrast, the O_2 trim system must be set to a fixed value and does not sense poor combustion, combustibles or formation of carbon monoxide.

Figure 10.11. Control system linearity. Non-linear response can cause control problems and inefficient operation.

Figure 10.12. Characterizable fuel oil valve.

Figure 10.13. Different delivery rates from air and fuel systems.

Figure 10.14. Over fire draft control system.

Figure 10.15. A fixed damper buffers high velocity escape of flue gas without limiting cross sectional area of stack.

Figure 10.16. Stack dampers used to block heat losses.

Figure 10.17. Oil-water emulsions, micron sized water droplets are mixed with fuel.

Figure 10.18. Flame retention head burner.

Figure 10.19. Heat recovery unit installed on a hydronic boiler.

Figure 10.20. Various types of firetube boilers with turbulators installed.

Figure 10.21. Automatic blowdown control eliminates high concentrations of impurities and waste of heat by excessive dumping of hot boiler water.

Figure 10.22. Increasing Cycles of Concentration.

Figure 10.23. Btu recovery from condensing flue gases.

Figure 10.24. Direct contact heat recuperation capturing latent heat from flue gases in the condensing range.

Figure 10.25. Direct contact condensation heat recovery tower unit with heat exchanger.

Figure 10.26. Submerged combustion water heating unit capable of efficiencies in the mid 90% range.

Figure 10.27. Bare pipe heat losses.

Figure 10.28. The effect of combustion chamber material on smoke.

Chapter 11   Boiler Plant Calculations

Appendix A

Figures

Appendix A

Figures

Appendix A

Figures

Chapter 15   Boiler Water Treatment

Figure 15.1. Blowdown is reduced dramatically when Cycles of Concentration is increased.
Figure 15.2. Mist formation in boiler water with high impurity levels. The volume above the water is steam, steam bubble is shown for illustration.
Figure 15.3. Loss of heat transfer efficiency with scale thickness.

Appendix B

Tables

Chapter 1 Optimizing Boiler Plants, Establishing the Ideal Scene.

Table 1.1. Maximum economically achievable efficiency levels for boilers between 10 and 250 million Btus per hour.
Table 1.2. Maximum attainable efficiency levels for boilers between 10 and 250 million Btus per hour.

Chapter 2 Boiler Plant Efficiency

Chapter 3 Distribution System Efficiency

Chapter 4 Energy Management Basics for Boilers and Systems

Table 4.1. Higher Heating Value (HHV) of various fuels.
Table 4.2. "Steam Tables", heat content of steam and water at various pressures.
Table 4.3. Thermal properties of water.
Table 4.4. Temperature change with boiler pressure.

Chapter 5 Efficiency Calculation Methods

Chapter 6 Combustion Analysis

Chapter 7 Flow Measurement

Chapter 8 Control of Boilers

Chapter 9 Boiler Tune Up

Table 9.1. Trouble shooting performance problems
Table 9.2 Boiler tune-up procedure.

Appendix B

Tables

Appendix B

Tables

Chapter 15   Boiler Water Treatment

Index

Index

Index

Index

Index

Index

Index

Index

Index

Index

Index